THE AISLES HAVE EYES

To Dan—
With
best
Wishes
—Joe

THE AISLES HAVE EYES

HOW RETAILERS TRACK YOUR SHOPPING, STRIP YOUR PRIVACY, AND DEFINE YOUR POWER

JOSEPH TUROW

Yale UNIVERSITY PRESS
NEW HAVEN & LONDON

Published with assistance from the foundation established in memory of Calvin Chapin of the Class of 1788, Yale College.

Yale University Press books may be purchased in quantity for educational, business, or promotional use. For information, please e-mail sales.press@yale.edu (U.S. office) or sales@yaleup.co.uk (U.K. office).

Set in Meridien and Futura type by IDS Infotech, Ltd.
Printed in the United States of America.

Library of Congress Control Number: 2016947155
ISBN 978-0-300-21219-8 (cloth : alk. paper)
ISBN 978-0-300-23469-5 (pbk.)

A catalogue record for this book is available from the British Library.

10 9 8 7 6 5 4 3 2 1

For Judy

CONTENTS

THE AISLES HAVE EYES

1 A FROG SLOWLY BOILED

It's said that a frog placed in a boiling pot of water will escape, but if the water is slowly heated the frog will habituate and eventually die. Although scientists dispute the accuracy of this statement, no one in the audience of marketers objected—or even said it was ethically inappropriate—when a digital-advertising executive used the image in an off-the-cuff suggestion of how marketers ought to treat people in physical stores. The occasion was a conference called *The Internet of Things: Shopping* that the online trade journal *MediaPost* convened in Manhattan during August 2015. The speaker, vice president of one of the world's largest interactive agencies, wasn't invoking the frog image because he wanted to kill shoppers; he was addressing the concern that people would push back against beacon surveillance. Department stores, supermarkets, pharmacies, discount chains, and other retailers have among them already placed hundreds of thousands of these inexpensive devices throughout their stores.[1] If shoppers carry the right apps on their smartphones and have the correct technology turned on, the beacons will alert the merchants and they can send the shoppers personalized coupons or other messages associated with the goods in a beacon's proximity. "We need to do a lot of hand-holding with our customers in the new environment," he urged. The goal had to be to treat shoppers like "a frog in a pot of boiling water": they had to be introduced to all the changes slowly so that they would come to consider them as a normal part of their lives.

Brandon Fischer, director of predictive insights at the influential GroupM Next consultancy, offered a view of the in-store surveillance future that seemed to embrace the frog comparison. In a keynote talk at the meeting, he predicted that by 2028 half of Americans (and by 2054 nearly all Americans) will carry in their bodies device implants that communicate with retailers as they walk down the aisles and inspect various items.[2] Based on how long you hold an item the retailer's computers will tell whether or not you like it. Other signals from the implant will indicate whether you are nervous or cautious when you look at the price of the product you are holding. The analysis may lead, Fischer suggested, to a conclusion that a discount on the product may reduce your nervousness and lead you to purchase it. His argument was a blunt, optimistic case for biometric monitoring in stores. And just as with the frog-invoking executive, no one in the room protested. No one wondered, either, whether in just thirteen years people would realistically consider this activity natural.

The attendees' nonchalance might seem strange, but the retailing business was changing so drastically and confusingly that such statements may well have seemed plausible. The message the marketers were hearing at the meeting and throughout their industry was that retailing is entering a new, hypercompetitive era with internet sellers. Brick-and-mortar merchants—the department stores, supermarkets, specialty stores, and chain stores that are still the center of the retailing universe—will succeed only if they figure out how to trace, quantify, profile, and discriminate among shoppers as never before. But for stores to survive this transition, shoppers will have to slowly learn to accept, even welcome, those eyes in the aisles as part of their natural environment—sort of like the frog in the pot of water.

This book is about how and why all of that is taking place, and the impact it is having on individuals as well as on society at large. I'll show how a new generation of merchants—Walmart, Target, Macy's, Stop & Shop, Safeway, Lord & Taylor, and many more—is working with a set of technology organizations to build a new future for physical retailing. They are reorganizing shopping to capture data about us through the very media we carry, even wear (such as a Fitbit). The goal is to routinely track us, store information about what we buy and when, and score us based on that and other information. Different forms of these activities are already under way in many stores. Retailers are using increasingly sophisticated electronic monitors, which show up first as experiments and then become ordinary elements of shopping. We typically feel their presence through personalized discounts often linked to our rewards cards. Depending on who you are and where you shop, you may already have experienced early versions of the following situations—scenarios some shoppers would consider wonderful:

- As you walk into an upscale department store, you may or may not realize that your phone signaled your arrival. The store cares because you belong to its loyalty program and have achieved high-value-customer status. Your presence is indicated to a store representative, whose tablet calls up your photo so she can recognize and greet you. The tablet also reveals which clothes you looked at on the store's website during the past week as well as the clothes you clicked on when you accessed the store's ads while visiting other websites during that period. Based on previous purchases and the information it has concerning your age, income, occupation, and family status, the store's computer predicts which of those garments you will buy. It also suggests matching accessories, again based on your website visits, previous purchases, and the special predictive sauce that mixes these behaviors with demographic information. When you complete your shopping and go to pay, you are pleasantly surprised to find that the computer is rewarding your loyalty in the form of a 20 percent discount on your purchases that day.

• You enter your local supermarket, with the store's app on your smartphone. The app instantly springs to life as you walk toward the first aisle. It retrieves your shopping history and loyalty score from the firm's computers and links them to the shopping list and Web coupons you had loaded on the app at home. The computer analyzes this information and concludes that this is a stocking-up visit rather than a drop-in for only a few items. Complex personalization formulas, which evaluate your shopping list and your location in the store, present you with ideas about what to buy, recipes based on what's in your cart, and discounts. The formulas also factor in information the supermarket has bought from data firms about your socioeconomic status, and assessments on where you are on various product buying cycles. Do your purchase patterns suggest you are at moderate risk for switching away from a particular shampoo? Does your increasingly less frequent purchase of a specific brand of diapers suggest a situation that can be countered with a $2 coupon (through a deal with the manufacturer), or is it likely your child is now potty-trained and so no longer needs them? Based on your shopping history, the formulas predict the extent of discounts that are needed to make you feel good about your shopping experience while also getting you to spend at least 10 percent more than you did the last time you came in to stock up.

• A similar scenario takes place in the big-box discount chain you visit often to buy household items. In addition to the information the supermarket used to send you messages and deals, this merchant has bought predictive data about your likes and dislikes based on the products you discuss on Facebook. The chain also bases its formulas for offering you discounts partly on an "influence" score it has bought from a company that evaluates the number of friends you have on social media and your degree of influence on them. For this trip you use the store's app on your phone to scan the products you want to buy, incorporating the personalized discounts as you go. Your loyalty status and checkout history give you the ultimate reward: by using your phone to scan your purchases, you bypass the long checkout lines and instead simply push a button at a station near the store's exit. There the retailer's computers also compare your scanned items with your purchase history and conclude you have not stolen anything. No one searches your bags (as sometimes can happen at stores that allow product scanning throughout the store as opposed to a single designated checkout location near the exit), so you're out superfast. No sweat.

Then again, depending on who you are and where you shop, you may already have experienced early versions of these personalized experiences many would consider unpleasant:

- No store representative greets you when you walk into the upscale department store, because your customer status doesn't warrant it even though you belong to the store's loyalty program. The store's computer knows from your shopping history and background that you typically purchase clothes with greatly reduced sale prices, and that you are likely to continue doing this. Representatives therefore prefer not to spend time with you, but you don't mind not being shepherded through the store; in fact, you rather like wandering alone. However, at checkout you see people in front of you joyfully surprised with 20 percent discounts, and you're envious. If you were given an extra 20 percent off that already on-sale sweater you were admiring, you might be able to justify buying it.

- You live in a lower-income neighborhood and typically rely on a local independent grocer for all your shopping. This merchant doesn't accept electronic coupons, and the paper ones you receive in the mail or encounter in the store don't match your needs terribly well. You occasionally do go to the chain supermarket in a different part of town, and you've found that your smartphone app does give you some relevant offers. It seems, though, that the supermarket's computer doesn't know enough about you to give you the various good deals that you hear other customers discussing with friends as they circulate among the aisles. In fact, at the checkout you notice some shoppers getting great deals on products you would like to try, but you have to pay full price.

Even a coupon for $2 off on one of those goods would be nice to have. You wonder if providing the app with more information about yourself would bring you better deals. But even if that would work, you don't know how to do it.

• You stop at a convenience store to pick up a few things. Once inside you notice a sign stating that the store's cameras use facial recognition technology to search databases for people who have criminal records. You actually have such a record, though it's several years in the past. The presence of the data-retrieval software may or may not explain why an employee seems to be following you as you move through the store. It also may or may not explain the curt treatment from the clerk when you go to the register to pay, as well as his prolonged examination of the $20 bill you offer as payment.

Before you feel too relaxed because you are the recipient of great service and a lot of coupons, realize that the behind-the-scenes tracking may well have consequences you might not like. Retailers might hire statistical consultants to generate reports about your eating habits based on the food you buy or about your weight based on the clothes you look at online and in the store, or to develop more specific health prognostications based on the groceries and nonprescription drugs you purchase. Their portrait of you may result in some nice coupons for you to redeem now, but it may turn sour later as you age, as statistical formulas may well make unflattering inferences about you and your family. Consider, too, that some retailers sell or trade the information they compile about their customers; some even assign "attractiveness" scores to shoppers based on this data. The scores and the many points of information about you may affect the sorts of insurance offers, food and clothing advertisements, and travel deals you receive. And in the not-too-distant future the knowledge that companies have developed about you may lead news organizations to highlight, and even rewrite, certain stories for you, and advertisers to provide you access to certain pay television programs but not to others. Much of this will be happening—and so much is already happening—without your consent or knowledge.

Oddly, although these practices relate to the ongoing and widespread public discussion about privacy—many government hearings and papers, advocacy-organization reports, academic meetings, and popular press pieces discuss the ways online marketers and government agencies track citizens—retailers only barely figure in the debate. The shopping aisle has, in fact, received almost no attention even among academics who focus on the social implications of consumer surveillance—an unfortunate trend, because the traffic that retailers can track through those physical doors is huge. According to the Food Marketing Institute, in 2015 Americans made an average of 1.5 trips per week to a supermarket.[3] The National Association of Convenience Stores (NACS) reports that in 2014 customers made nearly 160 million visits per day to the 152,794 convenience stores in the United States—58 billion visits per year.[4] And according to the leading retail analysis company ShopperTrak, during November and December 2013 Americans paid 17.6 billion visits to malls, department stores, "big-box" stores such as Walmart and Target, and specialty retailers such as Express.[5]

Although visits to supermarkets and convenience stores have remained rather steady in recent years, foot traffic to the stores that ShopperTrak audits has decreased substantially; the 17.6 billion in 2013 had been 33 billion in 2010.[6] Industry insiders generally believe that this drop is the result of increasing numbers of purchases over the internet. Because people can now shop electronically and have access to quick-delivery options, physical stores are competing with sellers not just in the same city or country, but from all over the United States, and even the world. What's more, even when shopping in a physical store, customers access the internet from their smartphones to use as a competitive weapon: product ratings, price comparisons, comments of friends via social media, and ads from competitors all affect whether and how much people buy. For their part, most brick-and-mortar merchants have tied part of their fortunes to electronic sales, and in the process now can also successfully track and profile shoppers, largely without their permission. Yet despite the huge growth of online commerce in the past decade, numerous studies indicate that over 90 percent of product purchases in 2015 still occurred at checkout counters in physical stores, and few in the industry are close to suggesting that they will fade away in the near future.[7] Industry experts do agree that brick-and-mortar merchants will succeed only by making the tracking abilities of their physical stores at least as good as or better than the virtual ones for targeting individuals with products and pricing. This includes using their data sources to create personalized messages for shoppers as they track them entering the store and proceeding through the aisles.

The implications are profound. The retail industry's data-centered activities are restructuring the architecture of both physical and digital retailing as well as the relationship between the two in ways that turn enormous information gathering into something customers take for granted. To make shoppers they care about feel good about making purchases, merchants are fashioning new visions of "rewards" that remake the retail phrase "owning the customer" for the internet age. "This is an era of unprecedented change for retail," Target's chief financial officer told the *New York Times* in 2014. "In order to win," he said, stores must keep "guests engaged with you as a business."[8] For Target and many other merchants, building relationships with individual shoppers today requires monitoring and discrimination. Retail monitoring involves gathering or purchasing information about shoppers' backgrounds and activities with or without their explicit permission or knowledge. Retail discrimination has two meanings, numerical and prejudicial. In the first sense, discrimination is a human and organizational impulse to note differences among things and among people. In stores it means maintaining records on individuals and performing complex quantitative analyses on that data—for example, determining which incentives will get particular individuals to buy more goods. Prejudicial discrimination follows as a result of the value the retailer places on each individual based on perceived differences: the retailer will offer specific shoppers different discounts on particular products reflecting the data's statistical portraits of those individuals and their families.

Part of retail discrimination is to identify customers deemed to have a relatively high "lifetime value" (a shopper's lifetime being typically defined as five years). These profitable shoppers receive tailored deals aimed to keep them coming back. Customers on the lower-valued end of the shopping spectrum typically aren't treated poorly; they may even get personalized discount offers in the hope that their value to the store might increase. But they will not enjoy anything like the attention and value the loyal customers enjoy. Moreover, some retailers downgrade the benefits of their loyalty program for customers judged to be of less value to the store based on the amounts they spend.[9]

These monitoring and discrimination activities usually fly under the public's radar. An exception was a *New York Times* article discussing Target's analysis of customer purchase patterns to identify pregnant customers. In the case of one family a father discovered his teenage daughter's pregnancy only because the retailer had mailed her a package of pregnancy-related deals.[10] These occasional press accounts no doubt cause at least some concern among those who also worry about contemporary government and corporate surveillance, but this coverage ignores a deeper development. The transformation of retailing is, at the core, a rethinking of the ways merchant-customer relationships should take place in the twenty-first century. Monitoring and discrimination are certainly critical to this rethinking, but they are only part of an accelerating project to redefine relationships with shoppers. Retailers' strategy is to mix shrewd loyalty programs, high-tech tracking instruments, and esoteric statistical manipulation with soothing brand images and smoke screens in such a way that shoppers accept systematic biases about them. Merchants ramp up "rewards" regimes (which they internally call "loyalty programs") that position store monitoring and discrimination as useful twenty-first-century values. In the process, the retail industry develops ever-evolving technologies to identify desirable prospects, encourage and maintain customer loyalty, and personalize attractive deals for individuals declared winners by the algorithms—and less-attractive deals for algorithmic losers.

We are only at the beginning of this new road. Stores are still struggling with how, when, and where to implement their data-driven regimes. The reshaping of retailing will likely add data-driven anxieties to both buying and selling. Stores will become centers of stress as dueling shopper and retailer technologies reach sometimes diverging conclusions about price and preferential treatment. Sellers will have to change prices constantly; introduce new products rapidly; and continually adopt new ways to define, identify, track, reevaluate, and keep wanted customers. Shoppers, who will continue to experience traditional stress about product quality and cost, will also now feel uneasy about what particular stores know about them and how the stores score them, and concerned about the impact this information will have on their shopping experience. They will also learn that trying to get on the good side of their favorite merchants means opening spigots to their personal information.

As of this writing policy experts, advocates, executives, and academics are arguing fiercely about the legality and ethics of data mining by online advertisers and the National Security Agency. Trade magazines, general magazines, newspapers, books, and a wide range of internet sites are sounding boards for many of these debates. In response, the Federal Trade Commission, Federal Communications Commission, Congress, the White House, and advertising industry groups propose and even push through regulations, laws, and self-regulation principles regarding concerns about social and individual harms. Ironically, at the same time, in a corner of society rarely mentioned by data policy makers, the retailing industry is working tirelessly to convince Americans to accept the notion that offering up information about themselves is a natural part of life. The very activities that dismay privacy and antidiscrimination advocates are already beginning to become everyday habits in American lives, and part of their cultural routines. Retailing is at the leading edge of a new *hidden curriculum* for American society—teaching people what they have to give up in order to get along in the twenty-first century.

Why is this happening, and how? And what do these changes mean for physical stores, the people who shop in them, and the larger society? The chapters that follow explore these questions, but first we should outline a social framework to help us understand what is taking place. Central to this framework are the meaning and the role of retailing's hidden curriculum—how it entered society, and how it spreads. The term *hidden curriculum* historically hasn't been applied to the world of merchants and customers. Philip Jackson coined it for his 1968 book *Life in Classrooms* to mean unintended lessons which are learned in school.[11] The basic idea goes back much earlier. The renowned turn-of-the-twentieth-century sociologist Emile Durkheim wrote in his book *Moral Education* about the generalizations students inevitably stitch together as an unspoken behavior code. "There is a whole system of rules in the school that predetermine the child's conduct," Durkheim said. "He must come to class regularly, he must arrive at a specified time and with an appropriate bearing and attitude. He must not disrupt things in class. He must have learned his lessons, done his homework, and have done so reasonably well, etc. There are, therefore, a host of obligations that the child is required to shoulder. Together they constitute the discipline of the school. It is through the practice of school discipline that we can inculcate the spirit of discipline in the child."[12]

Using *hidden curriculum* as a term to encapsulate Durkheim's meaning, Jackson and others followed him in exploring under-the-hood concepts young children pick up as part of their schooling. Much of their writing centers on how the hidden curriculum connects young people to the structures of power in society and defines their relationships to them. Some analysts describe ways schools implicitly lay out norms and values that are crucial for navigating the outside world; these ideals teach students "an approach to living and an attitude toward learning."[13] Jean Anyon echoes French sociologist Pierre Bourdieu in stressing the importance of cultural capital in the kinds of hidden curricula students learn about how they should relate "to physical and symbolic capital, to authority, and to the process of work."[14] Similarly, Samuel Bowles and Herbert Gintis argue that a connection exists between the unstated lessons schools teach children on one side, and the social classes to which the children belong on the other. "Schools do different things to different children," they insist. "Boys and girls, blacks and whites, rich and poor are treated differently."[15] The result, they contend, is a silent message about the future paths they should take—for example, the low-income or high-income occupational choices that best fit them. Paul Willis adds that the hidden curriculum of schools stubbornly reproduces the existing power structure of the society; even the ways student cliques form and resist administration policies within schools often echo the class distinctions of the students.[16]

Of course, any kind of situation can teach unintended lessons. Moreover, the notion that an institution—for example, religion, medicine, the law, and retailing—tends to re-create the dominant economic and political structures of the society is at least as old as Karl Marx. What makes the concept of a hidden curriculum attractive is not so much that it stresses broad, mechanistic lines of capitalist social power. Rather, the idea refers to patterns that quietly encourage students to absorb and act out their present and future social roles through the repetition of rules, stories, and performances that reflect, sometimes inconsistently, on a range of social status levels.

George Gerbner broadened the context of the hidden curriculum when he used it in relation to popular culture in general. An undergraduate folklore major who received his PhD in education, Gerbner spent most of his professional life as a professor and dean of communication studies. He received federal grants to study the portrayal of teachers in movies and television during the late 1950s, and the nature of violent images on television during the 1960s and 1970s. Conducting this work through the lens of folklore studies, he understood that telling stories is fundamental to every society. No stories are objective reproductions of reality. Folklorists emphasize that tales of all sorts serve up systems of messages that parade norms, values, and moral definitions, and tell people how to frame the world. Gerbner noted that the key difference between a society's traditional stories and those told by television, movies, and magazines lies in folklore's "handicraft" nature. Created and diffused organically through society, folklore is a product of the individuals who tell the stories. Modern media tales, news and entertainment, are by contrast created by industries where the concerns of organizations take precedence over all else. The stories result from media firms' need to satisfy marketers, investors, and, sometimes, politicians who benefit from showing people certain views of reality and not others. In a 1972 article Gerbner goes on to apply the term *hidden curriculum* not just to the educational institution but to the education people receive via widespread media—particularly television—about all institutions from their depictions in news and entertainment. He writes:

> The facts of life in the symbolic world form patterns that I call the "hidden curriculum." . . . [It] is a lesson plan that nobody teaches but everyone learns. . . . Culture power is the ability to define the rules of the game of life that most members of a society will take for granted. That some will reject and others will come to oppose some of the rules, or the game itself, is obvious and may on occasion be important. But the most important thing to know is the nature and structure of the representations that most people will assume to be normal and inevitable.

Following Gerbner's logic, it's no great stretch to port the idea of the hidden curriculum to the retailing scene. For one thing, stores are cavalcades of media depictions reaching out to people with ideas about what's good, bad, and ugly. Images on underwear packaging suggest the ideal body. Photos of athletes on cereal boxes suggest the sports that society should consider important—and the idea that people, especially kids, should be interested in sports at all. Tables of data and testimonials on all sorts of food containers call out norms of health and how to get there. The puzzles on kids' snack boxes quietly define intelligence and measure where a child falls within this definition. These sorts of representations suggest certain ways of thinking about the world in line with various alleged fitness, health, sports, and education authorities. But stores and the malls in which they are often enclosed also provide learning opportunities about what social characteristics mean. Different settings bespeak different notions of style in relation to various social distinctions of gender, race, ethnicity, lifestyle, and, especially, age and income.

The display photos of people in store windows and around selling spaces are often the most obvious signals of who is an anticipated customer and who isn't. Senior citizens, for example, are not poster people for clothing chains that want to appeal to a more contemporary crowd. Outdoor recreation stores tend to depict mobile, active enthusiasts, not individuals who are confined to wheelchairs. A store that wants to attract a wealthier clientele typically places upscale decor at the entrance, has wider aisles, and outfits its clerks in more formal wear. By law anyone can enter any of these places and make a purchase. And in fact many do cross the boundaries, for example shopping at both Walmart and Tiffany, Target and Louis Vuitton. Yet mall operators understand the meanings of these distinctions quite well, and they train shoppers accordingly. Mall planners cluster stores based on age, wealth, and lifestyle; establishments aimed at young adults, the rich, discount seekers, and outdoor enthusiasts are among those that each get their respective areas, often with distinct architectural designs and signage. Americans may or may not agree with the features the retailing establishment uses to draw attention to particular goods, but they do not actively rebel against them. The prevalence of the symbol systems that designate trendy clothing, outdoor happiness, and wealth suggests that, to the contrary, many shoppers take them for granted and accept that they accurately reflect the way of the world.

Retailers did not intentionally create these distinctions any more than schools or television created their hidden curricula. Nevertheless, these hidden curricula serve an institutional purpose: to cultivate and maintain discipline in the case of schools; to present readily understandable elements for telling stories (and generating profits) in the case of popular media; and to tie goods and shopping to certain meanings (also to generate profits) in the case of retailing. Now we are on the cusp of a retailing era that is adding an entirely new layer of routine surveillance activities and that carries with it the accompanying underlying lesson that it is common sense for shoppers to accept individualized profiling and deal making as part of the process of buying things. The importance of accruing data to learn about individuals certainly did not originate with the world of shopping. Yet, as with schools and media, retailers felt a strong need for such activity. In the trade press, at trade conferences, and in more than two dozen interviews for this book, people associated with retailing stated the need to retrieve a tsunami of data from customers, and this information is changing the ways merchants think about their businesses and about the people who frequent them.

Moreover, recalling George Gerbner's view of the modern symbol system, the engines powering this new world with its new hidden curriculum go beyond the retailers and their traditional suppliers. True, Walmart has built its own audience-tracking-and-analysis division, but most other retailers are tying into a large and growing subindustry of firms that aim to be part of the steps that turn customer data into gold. Terry Kawaja of the investment banking firm Luma Partners has attempted to chart the players in the online and mobile app part of this new world. He calls his pictorialization the "Commerce Lumascape," encompassing hundreds of competing companies involved with retailing under labels such as "order management," "analytics," "white label solutions," "software as a service," and "price comparison." Kawaja has not, so far at least, created a chart that describes these sorts of firms in the physical retailing space. Yet they, together with their online counterparts and major retailers, are building monitoring into people's routine daily activities. The transformation of physical retailing as a result of this work is creating the new hidden curriculum for our time. The word *transformation* is important. The world of physical retailing is a radically mutating place. Neither Gerbner nor the writers who focused on schooling's hidden curriculum dealt with a part of society that was transforming radically. My goal is to show how the retailing institution creates a hidden curriculum as retailing re-creates itself.

Of course, not all of what people learn about how an institution works is unstated. Some of it, such as not to shoplift, may be quite direct. But people do absorb many unstated messages from schools, the media, retailing, and other institutions about how the world works. They integrate them into everyday interactions with those and other institutions. This is not a straightforward process. We may be taught to believe that the ways we deal with stores stem from our individual needs, resources, and personalities. Yet we often don't consider the possibility that many within retailing often force, encourage, cajole, or otherwise get people in relationships with them to perform certain basic activities in ways that benefit their organizations. Doing that routinely leads shoppers to translate the organization's hidden curriculum into the "common sense" of everyday life. Sociologists Thomas Berger and Peter Luckman, authors of the classic book *The Social Construction of Reality,* wouldn't be surprised; it's simply how institutions work.[17] They note that institutions, "by the very fact of their existence, control human conduct by setting up predefined patterns of conduct, which channel it in one direction as against the many other directions that would theoretically be possible." They add that the institutional world requires "the development of specific mechanisms of social control" that encourage people to do things that support the continued existence of those institutions.[18]

Although it may not seem so at first glance, the creation of loyalty is a key mechanism of social control retailers try to make into common-sense. That is, through loyalty, merchants aim to lead customers to engage in everyday activities that strengthen the merchants' businesses. In a long and thoughtful treatise on loyalty, legal philosopher George Fletcher implies that you gain loyalty from others—that is, you cause them to stay with you—by offering two fundamental benefits, protection and privilege. We don't like to think of friendships, marriages, or citizenship in these terms. But people's loyalty to individuals and to countries involves a sense that the other person—or the country—will treat them in a special way compared to others who are not in that relationship. These bonds also bring with them an expectation that the relationship is reciprocal. The person receiving protection and privilege knows that he or she has responsibilities toward the other person, country, or organization offering those resources. That will involve offering protection and privilege in return. For example, in a marriage this would mean assistance in times of illness and exclusivity in sexual relations; in the case of a country it may be a decision to serve in the military or volunteer to help the government with expert advice during a war. Individuals will ascribe their own meaning to "privilege" and "protection," and the definition can also vary depending on the cultural norms of the time.

The same holds for retailing. Merchants often let shoppers know quite clearly that they can benefit from consistent purchases, though the definition of "consistent purchases" may vary by era and by individual store. Similarly, the retailer's repayment of that fidelity with privilege and protection can mean many things. *Privilege* can mean access to better prices, better or more relevant assortments of merchandise, advance knowledge of new merchandise, higher-quality items, gifts after a string of purchases, or superior credit terms. *Protection* might mean offers of security, safety, or other forms of help—for example, ensuring that a purchased tool works, a food is not contaminated, or the merchant's suppliers are careful. Exactly how loyalty plays out in any given era is the result of the retailing institution's hidden curriculum and ways particular retailers enact it by reinforcing shopper behaviors for certain everyday activities. It may also have to do with what the retailer sells, how it defines itself among its competitors, and whether it can afford to buck the institution's existing norms.

Now consider what is taking place in the new data-driven retailing system. The ways stores define a customer's loyalty and how they should reward it are increasingly dependent on complex analyses of the data retailers accumulate from tracking shoppers and buying additional information about them. One major change from the past is how retailers reward loyalty. Instead of being a straightforward tit for tat based on frequent visits, loyalty rewards have become tools that stores use to track customers, with or without their knowledge. Moreover, retailers and their consultants are increasingly applying loyalty as just one of many metrics to decide which announcements, deals, and discounts they should offer to certain individuals and not to others.

While mainstream retailers continue to encourage shoppers to consider loyalty a reward, it is actually giving way to complex algorithms that often punish people for fidelity. Indeed, a marketing executive of a major retailing chain notes that his company actually lowers prices for individuals deemed less loyal while keeping the prices higher for the ones identified as more loyal.[19] Other executives point to the different kinds of messages individuals receive from their companies based on the profiles the companies have gathered and the value they see in those profiles. Yet loyalty is only one element of new retailing regimes being organized to personalize messages in physical stores based on tracking and targeting shoppers virtually everywhere they go. As we will see, department stores, grocery stores, and discount chains are overturning more than a century of democratic impulses regarding loyalty. At the same time, shoppers are reinforced for following a new hidden curriculum that makes the giving up of data—and the bringing in of numerical and prejudicial discrimination—part of the shopping experience.[20] In the following pages I detail the behind-the-scenes activities that are changing our everyday shopping experiences in ways we as a society may not want. The activities represent a quantum technological and philosophical departure from how stores operated even at the end of the twentieth century. What we see now, though, is just the beginning. The institutional apparatus seems pretty well oiled, with determined executives and cutting edge technologies. This book follows this group as they create maps for a world that will redefine buyers' relations with sellers. More than that, the new retail system will encourage shoppers to accept a new, tougher reality in which the old saw that every person's dollar is worth the same increasingly no longer holds sway. And that means giving up on the historical ideal of egalitarian treatment in the American marketplace.

Today the shopper's phone stands at the heart of the change taking place. "A consumer essentially carries around a whole store on the smartphone in his or her pocket," notes a retailing consultant.[21] And, as improbable as it may seem, in a few years a chip implant in the shopper may also be part of the in-store messaging system, if technologists can convince enough people that the chips will help them be winners in the aisles. But first it is important to understand what started this transformation and why it's begun, as well as what is pushing it forward. As Berger and Luckman note, "Institutions always have a history of which they are products. It is impossible to understand an institution adequately without an understanding of the historical process in which it was produced."[22] That's certainly true about retailing, but it's a story not commonly told. The journey is outlined in the next chapter.

2 THE DISCRIMINATING MERCHANT

"When I first began my experience as a consumer," Christine Frederick told a congressional hearing in 1910, "I thought that the best way to do my family marketing was to ask the dealer his price and then Jew him down." Her anti-Semitic comment notwithstanding, Frederick was a famous household efficiency expert and consulting editor to the *Ladies Home Journal.* She learned, she said, that bargaining "was the attitude of all consumers in this country 50 years ago [in 1860]. This condition existed because at that time all dealers and merchants overpriced their articles, and the shrewd buyer was the only one who could get the best trade or bargain, after hours of talk and discussion." Price negotiating, she continued, "is still extant in some European countries and almost all Asiatic countries, where I myself have been first asked two dollars a yard for cloth and finally secured it at fifty cents. But [the American department-store merchants] John Wanamaker and Marshall Field saw the fallacy of these methods as affecting more important lines of merchandise. The reason why today 99% of all merchandise is no longer sold after these Asiatic methods is because common sense has demonstrated that the one price plan is more honest and more efficient for all concerned."[1]

The 1910s marked a period in retailing when it was not hard see how the old world of the previous decades was passing, but difficult to understand the new world that was being born. People knew things were changing dramatically, and tensions ran high. Christine Frederick was on Capitol Hill to support a bid by packaged-goods manufacturers to set their products' prices so that grocers could not lower them. Stores and brand name makers were struggling for new ways to exert primary influence over the shopper's purchases. All sides realized price would be only part of a solution that would encourage repeat purchases and satisfactory profits. Producers failed in their attempt to dictate the price by law, but they tried other forms of leverage over grocery stores and other retailers. The retailers, for their part, realized they had lost some influence over customers by abandoning "Asiatic methods" that involved one-to-one give-and-take. They tried to establish new ways to control their relationships with them in an era when the opportunities of face-to-face personalization were hard to find, if not lost.

To arrive at an understanding of the transformation of today's retail spaces, we first have to explore approaches to customer relationships, what they meant for the formation of retailing in the twentieth century, and the legacy they created for retailing in the twenty-first. I'll focus here on the development of two pace-setting shopping environments of the twentieth century, the department store and the supermarket. The time span cuts across three periods: the age of peddlers and small store merchants before the mid-nineteenth century; the rise, growth, and travails of department stores and supermarkets from the mid-nineteenth century to around 1995; and the birth of new forms and norms of selling with the rise of the Web, the smartphone, and new forms of industrial competition. Traditional pre-nineteenth-century selling techniques centered around prejudice and discrimination, with peddlers and small merchants determining which shoppers were "winners" and which were "losers" in the course of negotiating sales based on such measures as the purchaser's race, gender, ethnicity, and income. In the mid-nineteenth century, a new breed of merchants tried to encourage customers' return visits—what sellers typically meant by loyalty—in a populist way. They spread the word that a democratic era of shopping had arrived, but strong impulses toward discrimination remained. Despite their dominant egalitarian rhetoric, the largest, most populist sounding merchants accompanied these professed democratic ways with inequitable practices. Discrimination showed up in many elements of twentieth-century department store and grocery life: pricing, purchasing, packaging, branding, display, promotion, location, architecture, payment systems, and labor.

While major retailers were willing to slug it out over the lowest prices in the short term, they feared that perpetually guaranteeing the lowest possible prices was a recipe for insolvency. Instead, they tried to organize their operations in such a way that they could profitably offer beautiful surroundings, periodic discounts, and other populist incentives that would encourage loyalty among broad populations of shoppers. Ironically, though, attempts by the supermarket industry to reorganize during the painful economic conditions of the 1970s ultimately led it and, soon, all retailers to adopt a technology that could make their businesses more profitable by returning to the explicit discriminatory mindset of the peddler era. In the twenty-first century the new approach would make surveillance of shoppers a habit, and data-driven personalization a must.

Christine Frederick was correct that bargaining was a fact of life not too many decades before her congressional hearing. In fact, this form of retailing in the early United States was centuries old. Whether informal transactions or those conducted in more structured settings such as outdoor markets and small stores, buyers and sellers had seemingly always tussled over the price and quality of merchandise. As far back as biblical times buyers and sellers both inevitably took into account the characteristics and cues of the other as each side tried to gain the upper hand in the bargaining. The nature and the cost of the merchandise was therefore often tailored to the particular transaction. Historian Claire Holleran remarks that in ancient Rome haggling over price was an everyday activity. She also points out that how and where an individual negotiated exemplified that person's status: "The act and the space in which shopping took place were central to the construction of particular lifestyles and identities." That was as true in early modern England and Renaissance Italy as it was in old Rome. In Renaissance Italy the wealthy tended not to negotiate for goods in public. While they might examine luxury items in public, they would retire to their home with the merchant to negotiate the price.[2] But for most people the bargaining tended to be far less refined, as they typically purchased goods from screaming hawkers and peddlers trudging through the street. If they were lucky they might manage a quiet conversation about price and quality with a peddler moving from house to house in the village.

Bargaining is, of course, a two-way street. Just as the buyer cautiously had to pursue the best deal, the seller had to survive on his trading skills. As a reminder of the original prices they paid for the goods, as well as to recognize the goods if stolen, peddlers marked the back or bottom of their wares with unique symbols. They also followed a number of other strategies to maximize their returns on investment. Historian Laurence Fontaine, who studied the notebooks of peddlers in Europe from the seventeenth through the nineteenth centuries, found they covered a limited territory and so regularly returned to the same villages. Consequently, they developed ongoing relationships with individual customers and kept track of the deals made with them. They would also keep track of those customers' friends and relatives, as they might learn of the deals and could expect to pay similar amounts for similar goods. In one notebook from mid-nineteenth-century France, a peddler noted the women who bought his goods but also linked them to the more important aspects of the town: the men to whom they were related (for example, "Roubi's mother," or "Mourau's wife"). He recorded the occupations of those men and their family connections, as well as how they were addressed in the village and how they presented themselves during his conversation with them.

By researching the networks of their customers, notes Fontaine, peddlers slowly and "almost surreptitiously" became part of the village culture. Sellers and buyers negotiated prices "within the context of a personal relationship, and through the manoeuverings of trading and bargaining."[3] These bonds allowed the peddler to tamp down the inevitable tensions on both sides of the negotiation and to encourage customer loyalty built on protection: a reputation for honest representation of his goods' quality. The peddler could also build loyalty by implying to particular customers that he was giving them and their family especially low prices. Knowledge about the customers came in handy, too, when allowing sales on credit. Judging from Fontaine's peddler account books, a large percentage of the buyers did not pay in full when making a purchase. By allowing personalized deferred payments, peddlers solidified their relationships with customers. Each return to the village was an opportunity to recover a small amount of money they were owed as well as to offer a few more goods on credit—a built-in way to encourage loyalty. The peddler's challenge was to collect enough of the debt to enable him to repay *his* creditors while retaining enough to feed his family. This didn't always happen; peddlers often had to borrow money in their own villages with the expectation of recouping the cash during their upcoming travels.[4]

The peddling business migrated to North America with the flood of immigrants who poured in during the eighteenth and nineteenth centuries. Some peddlers who managed to amass a bit of cash settled in communities and established small dry goods emporia or grocery stores. Most of this group continued a version of the peddler tradition that was then called "personal service," whereby sellers helped customers choose from goods that were behind a counter or in a storage area. Writing about Chicago grocery stores at the turn of the twentieth century, historian Tracey Deutsch notes that customers developed "more personal relationships with their grocers than they might have with transient peddlers or nonlocal market stall sellers."[5] In addition to selling groceries, these merchants offered a range of other services including food preparation advice, letter writing or translating for non-English-speaking customers, and offering information about neighborhood activities. Grocers also learned much about their patrons' family situations, a topic that often came up when the customer asked for credit, as many did.

As with the peddlers, the relationship between shopkeeper and customer could be fraught for both sides. Deutsch points out that traditional approaches which led to "close personal relationships" and reasons for loyalty "did not guarantee smooth interactions in grocery stores, just as they did not in other retail formats."[6] Along with Fontaine, she found that advancing credit was crucial in the sellers' attempts to win customers, even as the sellers worried about bringing in enough money and the buyers worried about terms and their ability to pay. Pricing issues also caused angst. Customers typically assumed that store owners would give trusted clerks the code to the pricing symbols so they could bargain knowledgably. Despite efforts by the sellers to put customers at ease, shoppers worried whether they were paying more than other customers. Ethnic customers suspected that grocers who were of a different ethnic background overcharged them and sold them inferior products. African-Americans, who typically had little choice but to frequent stores owned by whites, were especially suspicious that the lack of transparency of price and quality caused them to receive poor treatment.[7]

During the mid and late nineteenth century, these strains opened the door for ambitious merchants to offer alternative techniques to make customers feel more comfortable and thereby gain their loyalty. The key change was posted pricing, which started as a stand-alone selling idea but became a crucial part of the new U.S. retail institution. Through much of the 1800s dry goods store owners had followed the peddlers' routine of personalizing charges for individual customers. Although Christine Frederick and others credited John Wanamaker and Marshall Field with making the change, the idea actually long preceded them. As far back as the seventeenth century Quakers believed it was morally abhorrent to charge different prices for the same item, and Quaker merchants Potter Palmer (an early partner of Marshall Field), Roland Macy, and the founders of Strawbridge & Clothier followed this philosophy in the early nineteenth century.[8] By the 1840s standardized pricing as a means to encourage customer trust was evident. An undated business card of A. T. Stewart and Company, possibly from as early as 1827 or as late as 1841, described the firm's prices as "regular and uniform." Adam Gimbel, founder of the iconic department store, guaranteed fixed prices at his Vincennes, Indiana, trading post in 1840. Several New York stores were advertising one price as early as 1842.[9]

For the merchant, part of the attraction of fixed prices was that clerks didn't have to be taught how to bargain. This model became increasingly important beginning in the 1860s, as U.S. cities saw the growth of multistory, multidepartment emporia employing many people. A famous precedent was Aristede Boucicaut's Le Bon Marché in Paris. The store's name means *cheap,* or *a bargain.* Boucicaut founded it in 1838 to sell piece goods in a poor neighborhood but branched out into various types of women's clothing and beyond; by the early 1850s it had morphed into what we call today a department store. Le Bon Marché marked all of its goods with fixed prices and even offered a money-back guarantee. These policies influenced American entrepreneurs, who moved from selling single stock products such as bolts of calico (the original "dry goods") to a wide range of merchandise types in distinctively separate departments. The post–Civil War era was ripe for this retailing model. The growing urban population and an improving standard of living meant concentrated numbers of people with money to spend. The wide circulation of newspapers, handbills, and billboards in dense neighborhoods enabled the stores to advertise to large numbers of people. Mass transit in the form of horse-drawn trolleys could bring those residents to central shopping districts. At the same time, rabid competition among many small stores selling the same limited range of products yielded virtually no profits. As retailing historian Robert Hendrickson notes, "Farsighted merchants cast about for new things to sell, realizing that if they wanted to grow, they would have to offer a fuller line of merchandise."[10]

The model these forward-thinking store owners had instituted in the mid-nineteenth century, especially regarding their approaches to gaining steady customers, fed into an entirely new approach that would develop in the following decades. Historians of late nineteenth-century America agree the era was transformative when it came to selling goods to the American masses. William Leach describes what was happening as "the democratization of desire" and traces it to the 1880s. This was a time of rapid industrialization and the creation of a new form of market capitalism based around industries that focused on "the production of more and more commodities." By the late 1890s so much merchandise was flowing out of factories and into stores that businessmen feared overproduction, which might push prices so low they would drive manufacturers and retailers out of business. In addition, while real incomes of many Americans were growing, wealth was lopsided; a small number of captains of industry controlled much of the nation's assets. Advocates of the new market capitalism may have wanted to draw public attention away from criticism that the far-reaching resources of a relative few were hindering democracy by diminishing the political power of the many. To do that, they reimagined democracy as the right of each American to dream about achieving personal happiness by buying things. "This highly individualistic conception of democracy," writes Leach, "emphasized self-pleasure and self-fulfillment over community or civic well-being. . . . The concept had two sides. First, it stressed the diffusion of comfort and prosperity not merely as part of the American experience as heretofore, but instead as its centerpiece. . . . Second, the new conception included the democratizing of desire, or, more precisely, equal rights to desire the same goods and to enter the same world of comfort and luxury."[11]

The key, merchants learned, was to follow a strategy of low markup and rapid turnover that contemporaries attributed to Boucicaut. The idea was to be proactive about low-cost selling: purchase large quantities of goods for cash (thereby getting the supplier's lowest price) and sell them quickly at low margins. The incoming profits would finance the purchase of yet more goods for cash that also would be unloaded quickly. The challenge was huge, however. Sellers realized that a store's "turn" of its products was key. Success would come only if product turns could take place continually and on a large scale. These conditions pointed to a store with several departments and many different items.[12] They also indicated the need for a continual flow of large numbers of people. To achieve this merchants built on the trend of posted low prices, even though they recognized that competition primarily over price could lead to unsustainably low profit margins, even severe losses. The emerging department stores therefore went out of their way to cultivate loyalty through egalitarian versions of protection and privilege.

Protection came in the guise of the customer's right to touch the goods as well as guaranteed satisfaction—the stores accepted returns without argument or penalty. Department stores offered a new, egalitarian form of privilege by creating a beautiful environment to which all shoppers would want to return and in which they could feel special. The no-haggle returns and the marvelous surroundings were geared particularly toward women, whom the new merchants saw as their primary customers. Historians point out that the nineteenth century in both the United States and Europe was marked by a Victorian sensibility regarding women in public places. "Disreputable women were associated with the immorality of public life in the city," writes historian Mica Nava. "Respectable and virtuous women were connected to the home."[13] Such concerns didn't mean middle-class and wealthier women fully shunned urban streets without escorts. Chroniclers of the era have pointed to the large numbers of women who moved freely around the landscape to carry out philanthropic work.[14] Barbara Olsen argues that in some areas "American women by the mid-1850s could now shop alone, make their own purchase decisions from expanding product categories, and spend their own money earned outside the home."[15] Moreover, Nava notes, the "Victorian ideal of separate spheres" unraveled rather quickly in the urban landscape around the turn of the twentieth century as women engaged in the public world of education and the suffrage movement. These changes also tied into the loosening of Victorian sexual mores. Relatively mainstream notions around "choice of partner, courtship patterns, and independence of movement" mixed with more radical ideas about " 'free love' and sexual pleasure as an entitlement for women as well as men." Women's magazines and, later, movies spread these ideas to the distaff population at large. None of these changes came easily; literary and other depictions of the "proper female sphere" suggest the social tensions. But by the early twentieth century increasing numbers of American women were "making themselves at home in the maelstrom of public life . . . however contradictory, painful and uneven the process."[16] Department stores helped the process along by positioning themselves as oases of populist privilege—and havens of safety—in a nonfeminine urban world.

From their inception the stores "provided a particularly welcoming space for women."[17] To elicit the sense of special comfort, they outdid one another with elegance and showmanship in order to attract crowds. Again, Bon Marché had pointed the way. A. T. Stewart's 1848 Marble Drygoods Palace at Broadway and Chambers Street in Manhattan, though not yet a department store, demonstrated how to draw crowds in a U.S. urban center. The block-long edifice became an instant tourist attraction. The canny Stewart lured visitors inside not just to ogle at the architecture but to attend fashion shows, scan and purchase the high-quality European women's merchandise, and scoop up the great deals he offered through "fire sales" of distressed and damaged goods. Competitors followed suit, first via the elegant Ladies Mile that grew on Broadway around Stewart's and later uptown in fabulously appointed stores such as Macy's, Siegel-Cooper, Gimbels, and Bergdorf Goodman. Among the other emporia that arose in this tradition were Abraham and Strauss in Brooklyn; John Wanamaker and Strawbridge & Clothier in Philadelphia; Marshall Field and Carson Pirie Scott in Chicago; L. S. Ayres in Indianapolis; Lazarus in Cincinnati; and Dayton's in Minneapolis. Designers and architects appointed the edifices with large display windows, carved wood, polished stone, imposing mirrors, fancy elevators, and streamlined escalators. The stores were among the first public spaces to use central heating and electric lighting both for illumination and to make artistic impressions. Department managers laid out goods in gorgeous displays, and the mundane administrative functions were situated in areas the public would not encounter.

The French author Emile Zola described these developments (also taking place in Europe) as agents for "democratizing luxury." He also saw them as enticements for the broad middle class to spend large, often too large, amounts of money.[18] The edifices were typically situated on major thoroughfares easily accessible by mass transit. In New York City and Chicago, some of the stores even maneuvered to have subway or elevated line stops at their locations. While the lavishness drew tourists, the real intention behind all the glitter was to build a loyal clientele, which for merchants meant people who would continually return to buy items. Accounts of the period indicate that such customers would visit the same store frequently, even several times a week.[19]

With the influx of crowds of shoppers, the new mass-marketing entrepreneurs understood they were no longer able to tailor prices to the individual.[20] Although those shoppers who considered themselves winners in such give-and-take transactions might balk, the retailers were counting on a much larger population being happy with the new process. Store management spread the word that loyalty-generating practices were to be enacted in a populist, routinized manner. A. T. Stewart supposedly boiled down the notion to a blunt statement to his clerks: "Never cheat a customer even if you can." He also instructed them on the importance of gifts for customers who paid cash rather than wanting credit. "If she pays the full figure [does not ask for credit], present her with a hair of dress braid, a card of buttons, a pair of shoestrings. You must make her happy so she will come back again."[21] John Wanamaker, who set the bar high for pursuing customer loyalty, laid the general precept out in a more philosophical tone. "The chief profit a wise man makes on his sales," he wrote in 1918 in a memorandum book, "is not in dollars and cents but in serving his customers."[22] Shoppers could rest assured that a posted price (Wanamaker himself supposedly invented the price *tag*) along with a satisfaction guarantee meant they were not being taken for a ride.

But Wanamaker and other department-store entrepreneurs also aimed to create a service atmosphere that would cause their targeted (women) customers to see the stores as multifaceted environments that surrounded them with a sense of feeling special. To that end, clerks and floor workers were trained to be attentive to customer needs.[23] Installment purchasing and free delivery were among the widely available services. Most dramatically, the stores made actual shopping only one part of the retailing experience. Executives also attempted to cultivate loyalty by attracting people for reasons beyond shopping. For example, one could simply rendezvous with friends at a beautiful fountain or impressive sculpture in the center of the store. The stores presented concerts on-site; Wanamaker's turn-of-the-century store had one of the largest pipe organs in the world. The merchants offered special toy exhibits for children, particularly around Christmas time. They offered dining by establishing restaurants in the stores. Or, instead of going anywhere near the store, customers could phone in an order from home. Wanamaker was a leader here, too: as early as 1900 the company employed hundreds of phone operators to take customer orders day and night.[24]

Like department stores, grocers developed an emphasis on loyalty through egalitarian access as well as populist protection and privilege, though in somewhat different ways and, when it came to architectural opulence, on a slower timeline. The early leader was the Great Atlantic and Pacific Tea Company. Founded in 1859 with a small number of New York tea and coffee stores, by 1880 it expanded to some one hundred branches. It also added sugar to its merchandise, a decision that began its march toward becoming the nation's first grocery chain. Apart from fixed low prices, A&P units were not much different from the neighborhood grocery stores with which they competed. They offered everyone discounts and gifts for continued patronage. They also advanced customer credit and allowed free delivery, activities that were expected in the grocery business at the turn of the twentieth century. But in the face of a national debate about high food prices during the 1912 presidential campaign, some of the company's outlets abandoned those inducements. Instead, they attempted to gain loyalty by trumpeting efficiency-driven cost reductions. The idea was to keep grocery prices as low as possible. Implementation meant reducing the staff to a manager and a clerk who retrieved the goods for customers. A&P's populist spin was that the no-frills style protected its customers from the scourge of high prices. The strategy proved highly successful, leading A&P and other grocery chains to expand the concept widely. It was a rather Spartan paradigm; an A&P store was no John Wanamaker.

The next couple of decades saw two additional iterations of the no-frills grocery store—the self-service market and the supermarket. The first innovation came from Clarence Saunders' Memphis, Tennessee, Piggly Wiggly store in 1916. The place was rather small compared to most groceries of the day. His idea was to have the customers pick up the goods in the store themselves (instead of asking a clerk for assistance) and then pay at a central checkout area. Saunders knew this strategy would enable him to cut down on labor costs. He was also convinced customers would buy more if they could see and touch all the merchandise themselves. He therefore created aisles that facilitated customers' handling of the goods, and he provided baskets in which shoppers could collect the items they wanted to buy.[25] As with department stores, the public spin on the setup was one of democratic privilege. As early as 1922, the Piggly Wiggly chain boasted that its self-service model "fosters the spirit of independence—the soul of democratic institutions, teaching men, women, and children to do for themselves."[26]

The idea caught on. Kroger, Grand Union, Acme, and the many other growing chains adopted this model, and eventually A&P followed suit as well. The low labor costs associated with the self-service model encouraged the expansion of grocery chains, and the larger ones regionalized. Another factor influencing the chains' growth was their continuing power struggle with their suppliers over issues that had brought Christine Frederick to Capitol Hill earlier in the century. The stakes were high. Many grocers preferred to choose the manufacturers whose products they put on their shelves, because they could bargain over wholesale prices and get good margins. Some manufacturers, however, tried to force grocers to carry their goods on their terms by going over the heads of grocers to their customers to position the manufacturers' products as special "brands." Packaged goods companies such as Sapolio Soap, Pepsodent, Kellogg, and Procter & Gamble launched an avalanche of advertising in popular newspapers and magazines to exhort people to buy items with unique names, trademarks, and package designs. The firms assured shoppers they would protect them from the uncertain quality of counterparts in barrels and bins and even in packages with unfamiliar persona and provenance. A Toasted Corn Flakes ad in *Munsey's Magazine* of the early 1900s noted that "the package of the genuine bears this signature [W. K. Kellogg]." It quoted an imperious young women telling a merchant who was offering her flakes from a different company, "Excuse me—I know what I want, and I want what I asked for—TOASTED CORN FLAKES—Good day."[27]

Manufacturers were pushing the right loyalty button. This was an era when grocery shoppers were looking for protection, as newspapers in the late nineteenth and early twentieth century were rife with stories of unscrupulous sellers putting dirty, dangerous, and even lethal materials into foods. People likely also remembered the "swill milk" scandal that affected New York and Brooklyn in the 1850s and 1860s. Diseased cows that had been fed distillery refuse were then literally milked to death.

Frank Leslie's Illustrated Weekly cover in 1858 featured a drawing of a dairy getting milk out of a sick cow on a hoist. (The liquid, the caption said, "made babies Tipsey and often sick.")[28] Ads that exhorted shoppers not to accept imitations played on these worries and effectively forced even large grocery chains to carry branded goods in the name of customer protection. Procter & Gamble warned grocers that they had better stock their new vegetable shortening Crisco—which, unlike the commonly used lard, would not spoil through bad refrigeration—because the advertising and public relations campaign the company was launching would be so extensive that their customers would come clamoring for it.

In addition to offering protection as a way to build customer loyalty, manufacturers also adopted campaigns and packaging that pointed shoppers to incentives included with the product inside the box or package. Manufacturers also offered discounts or gifts for collecting proofs of purchase from the product packaging.

The brand manufacturers' growing clout had a critical implication for the grocery industry, which sometimes was forced to accept smaller margins (and profits) on brand name products. But whether brand name item or not, the stores were committed to low prices because of competition and the need to encourage loyalty. Therefore, with such tight margins they had to sell very large numbers of goods to achieve acceptable profitability. During the 1920s many chains addressed the challenge by aggressively expanding their number of stores, which enabled them to purchase products from manufacturers at the lowest possible wholesale prices. They then implemented bouts of predatory retail pricing (undercutting the costs charged by competitors) on those lower-cost items, along with offering gifts with purchases, to encourage shopper loyalty and eliminate rivals in particular market areas, thereby gaining more leverage over suppliers. Once a store drove out competitors it could then increase prices a bit on the contested products.[29] Chains that successfully followed this strategy raised their market share: in 1920 grocery chains accounted for less than 3 percent of grocery sales, but by 1935 five chains captured about 25 percent.[30]

The chains' growing power notwithstanding, the Great Depression of the 1930s saw the emergence of a second new model for the grocery store, the supermarket. Originally spelled as two words—"super market"—it was conceived by entrepreneurs from outside the grocery establishment who pushed the low-price, no-frills paradigm to a new extreme in those hard economic times. The formula consisted of a massive store (often located in an inexpensive abandoned warehouse or factory), local advertising and promotional ballyhoo about products to be sold below cost, and huge numbers of customers noisily and chaotically hunting for bargains. One of the first of these "cavernous, ungainly stores" (as Tracey Deutsch describes them) was started by former Kroger executive Michael Cullen.[31] Called King Cullen, the store was located in an abandoned Long Island garage and used empty ginger ale cases for display tables.[32] While Cullen controlled all the merchandising in its outlets, some versions of the Depression supermarket leased space to different types of grocery retailers under the umbrella of a single name. Yet another iteration saw butchers, bakery owners, and other specialized grocery retailers join together to share the costs under one moniker, such as Big Bear stores, which opened in New York and New Jersey in the early 1930s.[33] Although these early supermarkets did not serve nearly as many people as the major grocery companies did, they often caused quite a stir in the neighborhoods they served. Twenty-thousand people entered the Family Market Basket on its opening day on Chicago's North Side; this supermarket was 50 percent larger than most of the bigger chain store units. The owner of another Chicago supermarket boasted that his opening-day crowds were so large he had to call the police.[34]

Executives at the established chains observed the phenomenon warily. Kroger executives worried about the dangers of appealing to loyalty through low prices instead of other aspects of protection and privilege. "The early supermarkets were barebones bargaining houses that emphasized price," Kroger company chroniclers noted, "while Kroger felt it owed its growth and strength to its insistence on quality"—presumably through its service and its company brands.[35] Although they were used to competing on price, sometimes very intensely, chain executives nevertheless believed that the ongoing extreme low-price promises, enormous size, and chaotic nature of supermarkets made them unstable ventures. They insisted long-term value lay in linking low, but not lowest, prices with a populist stress on a stylish experience and decorum in their stores. "The super-store in its present form may prove to be a 'depression baby,'" one industry person sniped.[36]

In fact, by the late 1930s supermarket owners began to modulate their emphasis on lowest price and shift to other forms of competition such as decoration with "a feminine touch," wheeled shopping carts instead of baskets, high-quality meats, and a new emphasis on service.[37] This new wave of classy treatment, available to all, seemed to result in higher profits in the face of the stores' traditionally low margins. Kroger, A&P, Jewel, and other grocery chains began to construct expansive stores with large parking lots to encourage shoppers to fill their trunks with groceries. The building rate and size accelerated after World War II, as many of the American middle class and upper-middle class began to relocate to the suburbs. Stores averaged nine thousand square feet by the late 1940s (far larger than the chain outlets of a decade earlier) and twenty-two thousand square feet by 1957.[38] They contained a vast range of advertised goods as well as store brands in an array of departments that sold fresh meats, salads, fish, and baked goods, as well as health and beauty supplies. Many of these areas were self-service, in keeping with the overall mindset from decades earlier.

The structures themselves looked quite different from the A&Ps, Piggly Wigglys, and Big Bears of earlier in the century. In a 1951 article *Collier's,* a popular weekly magazine, celebrated the change with language that evoked the air of privilege suffusing the great department stores of the era. "Low prices are no longer the supermarket's only attraction," the article noted. The transformation was "the prodigious issue of a marriage between brilliant showmanship and the world's most modern distribution techniques." The result, it said, was a new world of efficiency linked to indulgence: "In a supermarket the housewife buys her groceries (and a growing variety of other things) faster, because of self-service, and cheaper, because volume sales enabled the stores to keep prices at a minimum." The article went on to enthuse: "Nothing that could conceivably lure the housewife has been left undone by the supermarket operators. Their grocery stores are the world's most beautiful. They've gone into color therapy to rest the shopper's eyes; installed benches to rest her feet; put up playgrounds and nurseries to care for her children; invented basket carts with fingertip control; revolutionized a packaging industry to make her mouth water; put on grand openings worthy of Hollywood premieres. They've completely made over the nation's greatest business—food—to attract more and more of her interest and her dollars."[39]

Like the grocery chains, many department stores also relocated to the suburbs. While the branches did not typically match the glory of the downtown edifices (which sometimes suffered because of suburban mall competition), their executives continued to stress that the recipe for shopper loyalty had to involve alluring architecture, display, and service.[40] Some chains, such as Sears and J.C. Penney, aimed for the mass market with low prices. Others—for example, Macy's and Lord & Taylor—presented somewhat higher priced goods. Still other department stores, Saks Fifth Avenue and Neiman Marcus among them—presented a wealthy image with very expensive items. A 1974 overview of department stores marveled at the Woodfield, Illinois, shopping center, then the largest in the world. Along with Sears, Marshall Field's, J.C. Penney, and Lord & Taylor, it had two hundred smaller stores, many of which were themselves part of chains. The authors of this write-up concluded that an attractive location was a good formula for nonprice competition: "It is a seven-day operation with shoppers driving many miles for the Sunday 'champagne brunch' served by [Marshall] Field's Seven Arches restaurant, one of more than 30 eating places in the big development."[41]

For two decades after World War II media reports referred to the purchase of food and everyday necessities in supermarkets as well as the purchase of clothes and home goods in great urban department stores as a wonderful sensory experience. Both retailing sectors understood the competitive usefulness of linking material desires with the very essence of American democracy. Both arenas ultimately aimed for shopper loyalty that was not primarily tied to price, and both echoed the rhetoric and the architecture of democratic access to privilege. They continued to promote self-service as a democratic, loyalty-inducing plus. They claimed that the very depersonalization of the new forms reduced the tension that had been at the core of merchant-shopper relations in the peddler model. Unlike the earlier model, in which the customer often felt scrutinized by the grocer—who might be judging purchases, offering substandard products from under the counter, or charging more because of ethnicity, race, or perceived low income—they promoted the idea that twentieth-century shopping furthered the democratic ideal of allowing (to quote William Leach) "everybody—children as well as adults, men and women, black and white—[to] have the same right as individuals to desire, long for, and wish for whatever they pleased."[42] During the 1950s and 1960s, government and retailing officials pushed this equality-through-self-service-and-materialism argument as part of the defense of capitalism against communism in the Cold War, claiming that "because supermarkets lowered food prices, celebrated freedom of choice, and made customers feel that they were being treated equally, they reduced the appeal of communism and showcased the real value of American capitalism and free enterprise."[43] When Soviet leader Nikita Khrushchev came to the United States in 1959, American officials made sure that he visited a supermarket so they could present it as a symbol of their nation's superiority.[44]

The mid-twentieth-century buoyant rhetoric notwithstanding, critics have noted impulses toward discrimination coursing through the new retailing model. While department stores "eagerly accepted all dollars," in the words of historian Susan Porter Benson, they early on divided their clientele into two broad groups: the more affluent "carriage trade" and the poorer "mass," or "shawl," trade.[45] Stores declared that class divisions were simply the natural consequence of fundamentally different shopper desires. Articles in the retailing press contended that the shawl group saw fancy fixtures and other upscale appointments as indicators the establishment would charge them higher prices. In contrast to this group, noted a turn-of-the-twentieth-century commentator, "People of culture and refinement dislike crowds and crushes in stores [and want] to trade at a store where there is plenty of room and an abundance of air, with surroundings of an elegant, not to say aesthetic character."[46] Department-store architects and designers claimed to be taking to heart the differences between the two groups with the creation of bargain-price departments which sold lower-quality goods, located usually in store basements. As Porter Benson notes, "The bargain-section strategy allowed department stores to pursue simultaneously their strategy of 'trading up,' or seeking an even wealthier clientele, and their goal of increasing sales volume."[47]

This basic bias showed up on several levels, especially before World War II. Outdoor windows sometimes signaled the customer bifurcation: front windows might show prestige goods, while side windows might contain sale merchandise.[48] Inside, store personnel were taught to reflect a desire for an elite version of personalization in the midst of populist extravaganzas. People who clearly had a lot of disposable income received special consideration. Salespeople and doormen were encouraged to greet high-spending customers by name, for example, and even phone them when the store received new items that might be of interest. Preferential service could include free delivery. Well-off customers had their special requests dealt with more carefully and their returns accepted more graciously than those of other customers. Another side to personalization involved matching clerks with customer type. Personnel directors used a range of stereotypes—including age, race, and birth status—to assign salespeople (typically women) to different parts of the store based on the anticipated customers in each section. For example, older and native-born women were sent to higher-price departments, while younger and immigrant women worked in areas that sold less-expensive goods. Capping all of these prejudices was the allowance of a charge account to only the most select customers. During the first half of the twentieth century, charge accounts were the province of stores, not national credit firms such as Visa and MasterCard. To lessen the risk of default, and because they understood that well-off charge customers spent far more than cash ones, department stores made it clear this was a special privilege granted "only [to] wives of 'substantial citizens' and not those of you [from] working-class families."[49]

The discriminatory aura pervaded supermarkets as well. As early as 1926, chain grocers worried that they were reaching only working-class and lower-middle-class shoppers because they emphasized price over services such as helping customers retrieve items from around the store and stationing a clerk in the produce section to help customers choose, weigh, and wrap their fruits and vegetables. Christine Frederick, who didn't hide her anti-immigrant and anti-Semitic sentiments, stoked these concerns by reprimanding the chains as off-putting to busy women because of their abandonment of personal help and their long checkout lines. Industry observers warned that "low prices were not enough to keep customers because women—at least the sort of women so coveted by chains—wanted more."[50] Consequently, grocery chains began to move somewhat away from their vaunted standardization and toward at least partially retrieving the service component. Sometimes this meant hiring multilingual clerks to work in stores in immigrant neighborhoods. More commonly, though, it meant ensuring that stores in more desirable neighborhoods had service components. Class distinctions in service became even starker during the Depression, when supermarkets in working-class neighborhoods often offered lower grades of meat, cheaper brands of goods, and fewer services than they did in middle-class and upper-class neighborhoods. Chain store management also preferred to build new stores in wealthier neighborhoods, a practice that took on special vigor after World War II.

By the 1960s critics were citing studies that supermarkets in economically distressed (often African-American) neighborhoods were dirtier, limited in variety, and higher in price than in more well-off districts, and stocked substandard foods. The customers and clerks who shopped or worked at supermarkets in poor neighborhoods were certainly aware of the discriminatory patterns, but these conditions were rarely publicized except in the wake of major events such as the rioting that occurred during the Depression and World War II, and following the assassination of Martin Luther King Jr. in 1968. The latter insurrection served as an ironic counterpoint to the Cold War trumpeting of consumer democracy just a few years earlier. These angry outbursts led to short-term attempts by merchants and government officials to explore causes and offer solutions, but the policy suggestions tended to be rather bland. Tracey Deutsch notes that even reformers "treated the facts (that lower quality was being sold in poor neighborhoods in lesser quantities and at higher prices) as intransigent and inevitable." She adds that an article appearing in *New Republic* arrived at similar conclusions about supermarket inequalities but didn't think that government interventions could possibly be effective. The article did offer several ideas for addressing the problem, including the placement of home economists in local stores to provide advice, and busing shoppers to better stores in better neighborhoods.[51]

Supermarket operators' response to the obvious inequalities was to acknowledge them not as consequences of their prejudicial discrimination—a practice they publically condemned—but as unfortunate results of larger economic problems or as evils of discrimination unrelated to retailing. In 1968 a Philadelphia woman who had heard the claim that chain supermarkets sold goods of lower quality in black neighborhoods confronted the president of the National Association of Food Chains. According to a *Business Week* article, the woman said that the official "and his colleagues deny everything—and then they explain why it happens. Suburban stores, they say, are bigger than ghetto stores, with more parking space, roomier aisles, better displays, and greater store traffic. The implication to experienced merchandisers is plain: fresher product, greater variety. . . . By contrast, land in ghetto areas is often costly, and parking space is in short supply. The result: fewer stores, cramped, crowded, and offering few products."[52] Yet as retailing indignities continued to simmer within disadvantaged communities, the public relations arms of retailing firms continued their campaigns "to celebrate [the supermarkets'] hard work on behalf of women."[53]

Rather than actively confronting socially corrosive prejudices, supermarket and department-store management throughout the twentieth century focused on a different form of discrimination in the midst of their populist pursuits: they needed to find ways to keep their most profitable shoppers coming back. Posted pricing, low markups, and an impressive interior had rapidly become generic elements of the twentieth-century retailing environment. Yet identifying this group had become extremely difficult. For one thing, the stores now had so many customers it was hard to identify the ones who kept coming back. For another, loyalty and profitability didn't necessarily go hand in hand. A store could lose a lot of money enticing customers to be loyal; for example, earlier in the century department stores had lists of wealthy people who took advantage of their customer status by returning very expensive merchandise frequently, insisting on special deliveries, and piling up lots of credit. Just as frustrating, short of reviewing sales receipts, stores had difficulty determining what even good customers expected. Learning and responding to customer desires was an impressionistic project in the early twentieth century. Even in big stores, upper management and product buyers relied on department managers and salespeople for their opinions regarding what their desired customers wanted.

During this time many in the academic community and some businesspeople began proposing that retailers adopt a "scientific" approach to the issue. By the 1920s, stores were collecting a wide range of information from customer surveys,[54] and in 1933 market research pioneer Arthur Nielsen made deals with a representative sample of stores to audit the products they sold, and with the results of his study he launched his drug and retail store index; a year later he debuted an index of department store and food sales.[55] Historian Sarah Igo has noted, though, that the quantitative projects tended to measure what individuals actually purchased, not "what they desired."[56] Some stores did give their customers surveys to fill out about what they wanted from the store, and a few department stores recruited customers as advisers on issues concerning service and merchandise.[57] But overall retailers found it frustrating to study customer desires using the tools of science. A Filene's department-store executive sneered that retailers were "merchandis[ing] on opinions not on facts."[58] In her exploration of pre–World War II department-store management Susan Porter Benson concludes that "they had only the foggiest and most impressionistic sense of who their best customers were and what they wanted of the store."[59] And Bill Bishop, a longtime supermarket industry observer and founder of the Willard Bishop consulting firm, reflected in a 2014 interview that before 1990 supermarket managers strategized about their customers "more from gut feeling than anything."[60]

Rather than focusing on a "scientific" fix for ascertaining what customers want, executives turned to efficiency efforts in logistical operations and dealings with suppliers. University of Wisconsin professor Paul Nystrom wrote approvingly in his 1916 text on retail selling and store management that "business magazines have teemed with articles upon efficiency and scientific management."[61] Merchant Edward Filene led the way for department stores, developing a much-copied "scientific" model plan in the 1910s in which a store's sales patterns were assessed to determine the range of items in the store and the amount of stock for each.[62] Supermarkets, too, learned that making internal operations more efficient and putting price pressures on suppliers could widen margins more effectively than trying to build loyalty. They also discovered that profits could be made by charging manufacturers for special displays of their products and for in-store advertisements. These efforts at maximizing revenue made financial results respectable despite razor-thin margins on many of the products sold.

Store employees tended to see such efficiency efforts in opposition to the goal of loyalty because they believed their attention was then shifted away from customers and toward paperwork and backroom operations.[63] Management clearly felt a need for both. They insisted money saved through efficiency could yield money for loyalty-generating activities. And many executives believed that, like their efficiency programs, the best loyalty programs would be those that could be evaluated quantitatively. But measuring loyalty proved extremely difficult. For example, it was virtually impossible to calculate the extent to which high-profile, expensive interior designs and architecture brought in customers; their utility was taken for granted as important. Likewise, another loyalty effort, the signal events department stores staged for their cities, such as the Thanksgiving Day parades put on in New York by Macy's and in Philadelphia by Gimbel's, seemed to make good business sense, yet by their very nature they could not be analyzed in terms of specific sales and loyalty-building.

Consequently, throughout the twentieth century merchants focused on pursuing other methods whose results could be evaluated at least to some extent by norms of "scientific selling." The most enduring of these was advertising, which in the first half of the century meant marketing to wide publics. Merchants saw two main benefits: the possibility of building loyalty by circulating positive paid messages about the store in local newspapers; and the ability to check the effect on sales of an item featured in an advertisement. Although today it may be hard to understand how retailers could connect specific sales to their newspaper advertising expenditures, in the first part of the twentieth century retailers used a more direct-marketing approach. For example, when John Wanamaker advertised in Philadelphia papers that a special deal on umbrellas would take place the day after the ad appeared, he and his copywriter, John Powers, believed that umbrella receipts on the day of the sale would indicate the success of the ad. This view continued into the 1940s, as stores anticipated that once shoppers made the trip to the store to purchase the sale item, they might then purchase other things as well, and at full price.[64] Comparing the number of umbrellas sold on the day of the sale with the number sold on a typical day would give the merchant a sense of the power of the ad and of the newspaper. Placing the ad in various other local newspapers might help distinguish the factor that exerted the greatest influence on the sale, the ad itself or the vehicle in which it ran. And an examination of the receipts from those who bought umbrellas could indicate whether the advertised sale also resulted in sales of undiscounted merchandise. Still, the conclusions would hardly be definitive because of the many variables that could lead shoppers to make purchases. And the anonymous receipts in this generally cash-only era wouldn't signify whether ad-related purchases truly led to the best kind of loyalty—repeat purchases even when the customer encountered no discounted merchandise.

Grocery chains faced a similar dilemma. They also advertised special sales on particular products and could attribute at least some of any resulting revenue uptick to these circulars and newspaper announcements.[65] Discount coupons that were incorporated in an ad or sent to people's homes could be directly traced to the publication in which they had been printed if they carried a corresponding code. If they did, certain broad relationships between the advertising and resulting sales were quantifiable, such as determining the percentage of redeemed coupons for a particular product in a particular neighborhood. But there was much the numbers couldn't tell, such as whether the coupons encouraged repeat shopping and therefore the loyalty that retailers prized so highly. Meanwhile, brand goods manufacturers such as Procter & Gamble broadly circulated their own money-off coupons to encourage loyalty not to any particular chain but to the products. Sometimes, in fact, manufacturers circulated the coupons to get shoppers to demand grocers carry items that they otherwise didn't stock.[66] This activity stoked the long-simmering tension between the makers and the sellers, as did the payment amounts manufacturers offered stores for their trouble of collecting and returning the coupons; stores perennially complained the reimbursement was inadequate.

The trading stamp program offered the one loyalty vehicle that merchants knew *could* help them quantify repeat shopping. Trading stamps were introduced as a way to encourage a customer to make repeated shopping trips to the same store. The customer would be awarded stamps based on the total amount spent—typically one stamp for every ten cents paid. After accumulating a certain number of stamps the customer would receive a gift. Originating in the nineteenth century, trading stamps were different from discount coupons because sellers distributed them upon purchase. The first stores to dole out the stamps—beginning in England possibly during the 1880s and in the United States during the 1890s—were small department stores for exclusive use in their establishments.[67] They were different from small gifts that customers sometimes received with their purchase—say, a baker's dozen or the small thread A. T. Stuart encouraged his clerks to distribute—because they were individually worthless. This was controlled loyalty; customers both received, and were rewarded for, their stamps from the same store, and so repeat visits were easy to monitor. Tracking the amount of time it took for a customer to earn a reward would enable the merchant to note the success of the loyalty program across its customer base. The establishment could also note the value of a particular customer.

But this store-specific program didn't last. The situation began to get muddled in 1896 when the Sperry and Hutchinson Company became a wholesaler of what it called Green Stamps. The idea was simple and had a populist spin that invited everyone into the game: Sperry and Hutchinson sold large numbers of gummed stamps to retailers for a tiny fee per stamp. The retailer would then award customers with one stamp for every ten cents they spent. (Early on, to discourage credit sales, customers would receive stamps only if they paid promptly in cash.)[68] The customers were to paste the stamps into a specially designed booklet, which, when filled, they could redeem for various products ("premiums") at an S&H center. S&H would offer exclusivity to a particular type of merchant—a grocer, a plumber, an electrician—in designated localities. For example, a 1910 newspaper ad for Davidson's Cash Store in Phoenix, Arizona Territory, extolled the firm's "price and quality" and noted that "they are the only people in Phoenix who give S. & H. Green stamps with hardware."[69] S&H's pitch to these merchants was that stamp-collecting customers would give priority to retailers who offered the stamps. That may be, but because the booklets were likely filled with stamps awarded by any number of different vendors, tracking customer loyalty to a particular store via Green Stamps was impossible.

That didn't matter to shoppers, of course, who liked accumulating the same stamps from different establishments. The tiny revenue S&H made per stamp ballooned into a huge cash flow, which translated into substantial profits in two ways. First, S&H profited from the difference between the price it paid for a premium and the amount in stamps a customer had to redeem for the item. Second, S&H earned considerable profits in the form of interest from the large cash reserves it held in the fairly lengthy period between the time that stamps were sold to stores and when customers completed the process of collecting, saving, and redeeming them.[70]

S&H's success encouraged other companies to enter the trading stamp business, mostly with a regional focus. One fierce competitor was Gold Bond stamps, known at the time for offering a mink coat as an item that could be redeemed for stamps. But while loads of consumers gathered stamps eagerly, others found the activity preposterous. Some economists scoffed that the stamps simply encouraged price inflation, arguing that stores raised their prices to pay for the stamps. Some state legislators screamed that the stamp companies were actually anti-democratic. They said the programs were taking advantage of naïve consumers, who didn't realize that the value of gifts amounted to only about 2 percent of what they had spent. One anti-stamper described such programs as "prostitutions at their best and economic insanity at their worst." Dozens of states tried to ban trading stamps outright or impose taxes that would force them out of business.[71] Yet most of these initiatives failed, and the stamps endured—but not because retailing executives liked them. Retailers tended to see them as albatrosses and tried to find excuses that would be acceptable to their customers for getting rid of them. To them, store coupons or discounts were far more useful because the executives could control the timing, nature, and amount of the offers, and these tactics would provide them with at least some quantitative measure of success. With stamp programs, all stores knew was the total number they gave out, though occasionally they could quantify the success of special promotions, such as offering double stamps for a short period.

Trading stamps experienced rapid growth up until 1915, with department stores, mail-order houses, and many other sellers including grocery chains joining in. With the start of World War

I the business slumped, and this downturn lasted well into the Great Depression. Department stores experienced cash flow problems during the war and were among the first to stop using them, while grocery chains canceled their programs as they converted their units into economy outlets. Consequently, stamp programs between the two world wars became the preserve of small stores wanting as many loyalty arrows as possible in their competitive quivers. But the competition that accompanied the fast growth of supermarkets during the 1950s led many supermarket chains, which by then seemed interchangeable as well as impersonal, to offer trading stamps as a way to stand out from the others. One retail economist wrote that, in the 1950s, "the loss of individuality reinforced the supermarkets' image of a formal business, carrying the same brands as the other supermarkets, offering the same conveniences, and charging just as much. In less than two decades the supermarket had become a comfortable and commonplace store to the shopper; an enmeshed and exposed firm to the merchant."[72]

Some supermarkets, including Kroger and A&P, held out for a time, concerned that offering stamps could reduce their margins and therefore limit their flexibility to time other premium and discounts programs. But most ultimately concluded that stamps were an unstoppable rage, as the president of Kroger reflected to the *Wall Street Journal:* "We fought them by cutting prices; we gave away hosiery, dishes, and dolls. We used every gimmick known—and still the stamps stores took sales away from us. We couldn't fight them, so we joined them."[73] A&P, whose president called stamps "a drag on civilization," began dispersing them at some stores as well.[74]

The trading stamp craze continued through much of the 1960s. Berkshire Hathaway, the chief investment vehicle of Warren Buffett, began investing in Blue Chip Stamps in 1970 when the company had sales of $126 million, with sixty billion stamps licked. "When I was told that even certain brothels and mortuaries gave stamps to their patrons, I felt I finally found a sure thing," Buffett recalled in 2007. But Buffett was buying in at the end of a long ride. "From the day Charlie [his partner] and I stepped into the Blue Chip picture, the business went straight downhill," he acknowledged.[75] Discount stores and inflation were the main reasons that merchants began to jettison the sticky things. Big-box chains such as Shoppers' City, Target, and Kmart sprang up on a large scale in the 1960s and started diverting the profits from department stores and supermarkets. They challenged shopper loyalty to stamps by aiming price-cutting efforts at the most popular redemption center items and setting up grocery departments with lower prices than those offered by supermarkets.[76] The discount model also undoubtedly benefited from widespread concern regarding rapidly rising food prices. Various factors contributed to this dramatic increase, among them steadily increasing petroleum prices, a decrease in world grain production that caused feed grain and therefore meat prices to skyrocket, and a devaluation of the dollar. The impact was dramatic. The consumer price index rose 16 percent from 1967 through 1970, and 27 percent from 1970 to 1974. Food prices rose even more—15 percent from 1967 through 1970 and 41 percent from 1970 to 1974. Wholesale prices of farm products rose higher yet—an astonishing 69 percent from 1970 to 1974.[77] The United States had never before experienced such high rates of inflation during peacetime.[78]

In this tumultuous environment, the profits of retail food chains as a percentage of return on sales plummeted from an average of 1.2 percent in 1963–70 to an average of .7 percent in 1972–74.[79] Struggling to keep their companies afloat, executives ceased giving out trading stamps. Even companies that ran their own programs, such as Grand Union, stopped them during the 1970s.[80] They instead decided to emphasize discounts and coupons the retailers could control closely, and in amounts that would encourage customer visits. In the early 1970s, though, the overriding concern for retailers was less loyalty than it was dismally low profitability. While discount coupons also allowed merchants to track results at least a bit more precisely than stamps did, this amounted to small consolation at a time of gravely bad earnings. In such an environment it's hardly surprising that when supermarket executives were presented with an invention aimed at improving managerial efficiency—which also might potentially quantify shopper loyalty as never before—they leaped at the chance.

The new technology was the Universal Product Code scanning system: the small rectangle of black and white bars that gets swiped across a scanner connected to the checkout register—a feature that we take for granted on virtually every package we buy today. Incorporated into the bars is a unique code for the particular brand and type of item. The checkout scanner reads the product code—say on a sixteen-inch DiGiorno Rising Crust frozen pizza—and looks up the pizza's current price from a database in an on-site or central computer. Almost instantaneously, information regarding the sale and the store from which it was purchased is transmitted to the corporate office. Corporate buyers can then determine when they need to replenish the stock for that specific pizza in that specific store. As a result of this development, companies now had immediate, precise knowledge of what was (or wasn't) selling and when, and in which stores.

Despite its evident benefits, the UPC system was not instituted until the mid-1970s even though the basic technology had existed for decades. In 1948 Bernard Silver, a graduate student at the Drexel Institute of Technology (now Drexel University), overheard a local supermarket executive imploring a dean to develop an efficient means for creating codes for product data. The dean demurred, but Silver took up the challenge with another graduate student, Norman Woodland, and later that same year they conceived and patented a version of the modern barcode. They used light from a very hot bulb to reflect off printed lines and create patterns that could be read by a special sound-on-film tube. As one writer notes, "It worked, but it was too big, it was too hard, computers were still enormous and expensive, and lasers hadn't been invented yet."[81] A further stumbling block was the lack of an industry system to designate numbers or products. As two business economists note, "The UPC system would have been prohibitively expensive, perhaps technically impossible, to implement a decade earlier than it was." They added, though, that it was the supermarket industry's search for efficiency in the midst of an unprecedented financial challenge that pushed the project toward success. "The UPC was shaped as much by the challenging and volatile conditions of the food sector as it was by the forces of technology."[82]

Between the 1940s and the 1970s some manufacturers and retailers tried to establish their own product coding systems, but each was incompatible with any other system and so they were effectively useless. Amid grave concerns about rising inflation in the early 1970s, a group of grocery industry trade associations banded together to pursue a universal coding system, forming the Uniform Grocery Product Code Council, a committee of supermarket and packaged-goods executives aided by the consulting firm McKinsey & Company. IBM was chosen to develop the technology, and National Cash Register developed the actual scanner. The first UPC-marked item scanned at a retail checkout took place at a Marsh supermarket in Troy, Ohio, on June 26, 1974, at 8:01 a.m.: a ten-pack of Wrigley's Juicy Fruit chewing gum.[83]

There were doubters, but the project moved ahead rather quickly. By 1976, 75 percent of the items in a typical supermarket carried a UPC symbol, while installation of scanners in supermarkets took place more slowly. The code was soon considered to be "firmly established in the food industry."[84] At the same time, Kmart, Walmart, and other grocery-stocked big-box merchants joined the move to scanners. Moreover, nongrocery manufacturers were now approaching the Code Council for UPC symbols to place on their products. By 1982, food and beverage manufacturers no longer constituted most new UPC registrations.[85]

The Wrigley's gum crossing the scanner signified the beginning of a revolution throughout retailing. Stores adopted the device slowly at first (in 1976 *Business Week* published an article titled, "The Supermarket Scanner That Failed"), but through the 1980s, and spurred by the big-box merchants, supermarkets made the scanners standard features at checkout. The executives who switched over to this system had efficiency in mind primarily, and over time they certainly achieved that goal. In addition to speeding the checkout process, the scanners enabled stores to order goods more efficiently because they could immediately identify which products sold poorly and which sold well. And stores no longer had to stamp or otherwise label each individual item with pricing information for a clerk to read and enter into a cash register at checkout; instead they could affix a single price label for an item to the adjacent shelf so that customers would know the cost. Finally, because they now could have virtually instantaneous knowledge of the goods in their stores and how they were selling, retailers could stock far more products than in the past and with fewer stock-keeping woes—and they did.

More important, perhaps, the scanning system upended the power relationship between retailers and manufacturers. For the first time in a century, chain retailers now had leverage over manufacturers when it came to information about the items they sold. Neither Nielsen nor IRI (another retail auditing company) could hope to provide anywhere near the level of competitive detail that retailers' computers were now accumulating. With the new technology, store executives could know with unprecedented speed how a manufacturer's new product was selling, or whether a brand's advertising or couponing program seemed to be working—and all this information could be further broken down by store, by the extent of the success of the item or program, and by specific durations. These were important bits of information with which stores could bargain with manufacturers for product discounts, promotional funds, and "slotting fees"—payments for including a manufacturer's product on the retailer's shelves.

The scanner system also offered opportunities for efficiency to intersect with loyalty. Grocery executives recognized at the outset that the system could encourage shopper loyalty, reasoning that customers would appreciate getting through the checkout line more quickly. And although the idea wasn't implemented at the time, the executives saw that the checkout registers could also be used to print recipes related to specific purchases—reflecting an interest in using personalization as a loyalty motivator in ways that were previously impossible. From there it was no great leap to view the barcode scanner as a logical platform that retailers could use to track the purchases of all customers. Trading stamps seemed to offer that potential in the early twentieth century, but ultimately they weren't relevant to the mass-market, populist impulses of the period. In the late twentieth century the pressures on retailers pointed to a very different, nonpopulist mindset: a high-tech version of the discriminatory peddler era. As the 1970s drew to a close, the idea of tracking sales of individual shoppers by computer and profiling people based on what they bought had yet to take off. But the institutional challenges of the next decade would certainly move retailers in that direction.

3 TOWARD THE DATA-POWERED AISLE

During the second week of December 2011, Amazon encouraged shoppers to install the company's price-check app on their smartphones and then scan the UPC bars of items they were considering buying at local physical retailers. As an enticement, the company offered a 5 percent discount (up to $5) if the customer purchased the scanned item from Amazon (the deal applied only to certain products). Not only could the shopper benefit by paying less, but the app also provided the e-tailer with the location and the price of the product—critical information pertaining to its physical-store competitors throughout the country during the most popular shopping time of the year. With this information Amazon could then adjust its prices by geographic area to achieve optimal profit margins.[1]

To many merchants, the mix of barcode scanning, location checking, and a discount—on top of no sales tax—reinforced Amazon's image as a destructive pirate. Of particular concern was Amazon's encouragement of "showrooming," which is the practice of inspecting merchandise at a physical store but then making the actual purchase online. The publishing industry has little doubt that this tactic caused the downfall of many

U.S. bookstores. The author Scott Turow, president of the Authors Guild, decried the "bare knuckles" approach to retailing that Amazon's broader price-comparison project revealed.[2] Other critics focused on the possible consequences for small businesses. The *Sun* newspaper in Lowell, Massachusetts, editorialized that "the online behemoth is trying to put small business owners on the street with a Christmas sales promotion that is—quite frankly—unfair and un-American."[3] The *New York Times* drew a parallel between Amazon's activity and "when Walmarts open in small towns."[4] Similarly, a Canadian bookseller rejected the admiring portrayal of Amazon CEO Jeff Bezos, who transformed the way many purchase books, "as Steve Jobs of the book world. For me, he's the (Walmart founder) Sam Walton of the book world."[5]

Large retail chains, including those with online components, were also angry that their major internet competitor was encouraging huge numbers of people to exploit the physical stores they frequented. "Amazon has been at the implicit war with local brick-and-mortar stores" since its inception, wrote *New York Times* columnist Robb Mandelbaum. "Last week, the implicit seemingly became explicit."[6] Target signaled its irritation in a letter to suppliers that promised "what we aren't willing to do is let online-only retailers use our brick-and-mortar stores as a showroom for their products and undercut our prices."[7] The customer "has so many more available places to get products, and mobile technology and the Web have completely exposed pricing," said Al Sambar, a retail strategist at the consulting firm Kurt Salmon. "So the two best competitive levers retailers have are under attack."[8] Five months later Target stopped selling

Amazon's Kindle e-readers.[9] The company certainly knew that Amazon's app was not the only smartphone price-checking software, but officials were especially angry that a giant online merchant was gleefully poaching customers systematically from inside a competitor's physical store.

The year 2011 signified the beginning of a great transition in America's retailing institution, as a broad swath of retailing executives expressed consternation over Amazon's activities and their inability to combat them. Retailers finally opened their eyes to the mountains of data that now could be accessed because of the Universal Product Code, the internet, and other new technologies, and they started to rethink their entire approach to customers, merchandise, pricing, and the selling process. As a result, we are entering a new retailing era with swiftly growing impulses toward customer discrimination and ever-quickening movement away from the democratic ideal. Ironically, those who are setting the pace are borrowing industrial approaches from the past to bring back aspects of personalized selling that marked the days of peddlers and small stores. Amazon's price-check stunt proved to be the tipping point that moved physical retailers toward personalization on an industrial scale. The pre-Christmas hijacking, limited though it was, forced brick-and-mortar merchants to recognize that the online and the offline worlds were coming together inside their physical stores. If they didn't use their aisles themselves to target likely customers with messages driven by the detailed individual shopper information that they had compiled, their competitors would. Once again the UPC, or barcode, would play a major role, this time in the merger of physical and online retailer.

* * *

To understand how we arrived at this transition we need to take a closer look at retailing in the late twentieth century—and in particular one retailer that laid the groundwork for change before Amazon. The rise of Walmart is the starting point for exploring how competition, technology, and rhetoric about targeting the customer combined to set the stage for personalization-centered physical retailing. Walmart practically sneaked up as a powerhouse on the American retail industry. "The plain fact is that the market has failed to realize the impact that Wal-mart has made," said a retail analyst for Smith Barney, Harris Upham & Company in 1989. But, she quickly added, "What it has done has never been done before."[10] From 1977 to 1986 the discount chain increased its annual sales from $678 million to $12 billion, and its net income from $22 million to $450 million. And in 1988 its sales increased 34 percent, to $15.9 billion, and its net income soared 39 percent, to $627.6 million. A *New York Times* columnist wrote that Walmart is "unarguably the most dramatically successful retail company of the last twenty years. It has leaped out of the South to become the nation's third-largest retailer after Sears, Roebuck & Company and the K Mart Corporation and is now hurtling ahead with the possibility of overtaking each of them in sales in the next few years."[11]

Some speculated that because Walmart had "leaped out of the South" it was not taken seriously despite its surging income. Launched in 1962 by Sam Walton and his brother Bud (the store signs read "Wal-mart" for many years to indicate its founders), the chain systematically radiated "in concentric circles" from its first store in Rogers, Arkansas.[12] Even by 1988 the company operated stores in only twenty-four states, while the corporate base was in Bentonville, Arkansas, far from the traditional halls of retailing power in the Northeast and upper Midwest. Yet during that decade it was already expanding beyond its basic discount store format to include Sam's Wholesale Clubs and, beginning 1987, adding a grocery section to many of its new, expanded stores, which the company called "Walmart Supercenters."[13] By 1987 Sam Walton, with a net worth that *Forbes* magazine pegged at $8.5 billion, led the *Forbes* list of richest Americans by a huge margin.[14]

Walton and David Glass, his president and chief operating officer, attributed Walmart's success to its laser focus on the customer. "As for Wal-mart becoming a household word, we never think about things like that," Glass told a reporter in 1989. "What we want is for customers in our territories to think of us fondly."[15] A Morgan Stanley report on the company that year agreed: "Management is insisting that every customer be treated as a guest. . . . This unique attitude is producing a strong response from consumers since service with a smile is often a forgotten commodity in retailing."[16] This approach democratically included stationing "greeters" by the entrance to the store to help all customers find what they need. The company also tried to draw in customers by emphasizing that the chain was the place where they could "Buy American." These efforts notwithstanding, most observers agreed that what brought people to Walmart in the first place was its reputation for low prices.[17]

That reputation tended to be well deserved, and it created great fear among executives of competing stores. Individual merchants could not purchase items at the kind of scale that gave Walmart leverage to demand rock-bottom prices from manufacturers. But beyond low prices, another reason for Walmart's success was its absolute concentration on cost control throughout its supply chain. In part, this was simply a matter of survival, as in its early years Walmart could not get large wholesalers to deliver to its rural Southern locations. According to an Associated Press account, the company therefore "had to create its own distribution system, with its own trucks, its own direct dealings with manufacturers, and its own technologies."[18]

Sam Walton saw that technology could aid in making that process ultraefficient, and in the mid-1960s he began using computers to track the delivery and sale of inventory in each of his stores. This predated the UPC system, so at first all the items coming in and going out still had to be counted individually. By the early 1980s most packaged goods were labeled with bar codes, and in 1983 Walmart outfitted its warehouses and store delivery areas with the UPC system, including hand scanners so that employees could check in merchandise as it arrived.[19] At the same time, the company replaced standalone cash registers with scanning systems that relayed each individual sale to an on-site computer; the store then reported this information to Bentonville. In 1987 the company greatly speeded the process by setting up the largest corporate satellite system in the world, providing real-time data, voice, and video links to all Walmart stores. Individual sales information from every store was now delivered instantly to the company's headquarters.[20] Walmart made the process even more efficient when it announced that it would no longer buy from wholesalers, instead dealing directly with manufacturers, and in 1988 it proceeded to force these suppliers to link their computer systems with Walmart's.[21] As a result of this electronic data interchange (EDI) operation—probably the largest in the United States—paper and fax relationships virtually disappeared.[22] Instead, all invoices came directly into Walmart computers for quick processing, and Walmart computers instantaneously alerted suppliers' computers when an item needed to be replenished.

Writing about Walmart in 1994, a Canadian reporter marveled at the speed and efficiency at which the retail giant was able to restock its shelves. "Wal-Mart's check-out scanners feed information by satellite dish to the distribution centers, orders are made automatically and products are speeded to a loading dock by conveyors. In the United States, every Wal-Mart store is within a day's drive of one of about 20 Astrodome-sized distribution centres. The centres are serviced by a fleet of 14,000 trucks, the largest private fleet in the United States."[23] The reporter described how, using information provided by the company's computers, Walmart personnel at each distribution center immediately transferred every manufacturer delivery onto the appropriate Walmart truck for carriage to specific stores. No product had to be entered into warehouse inventory, and the speed of transfer "allow[ed] Wal-Mart to re-stock its stores at least twice a week and sometimes every day, compared to the industry norm of every two weeks."[24]

These impressive feats, generally recognized as far more accomplished than what Walmart's competitors could do, were just the tip of an iceberg of efficiency-centered activities Walmart used to wring costs out of products in its now 2,440 stores.[25] One ongoing common tactic was to compel suppliers to lower the cost of their products by lowering expenses. Indeed, there were reports that David Glass, who remained CEO until 1993, would even chastise manufacturers for using private jets to travel to Bentonville for meetings with Walmart officials, and for locating their corporate offices in expensive cities. "You don't tell Wal-Mart your price; Wal-Mart tells you," said an executive of the American Textile Manufacturing Institute in 1997.[26] There were also leaked accounts of Walmart pressuring its vendors to submit to especially low wholesale prices on everything from dresses to kitchen utensils to lawn mowers. Publicly executives were upbeat. "I went [to Bentonville] knowing we were going to get squeezed and wrung and twisted all in positive ways," the CEO of Liz Claiborne said in 2000 about the Russ clothing brand his firm designed exclusively for Walmart.[27] To keep costs extra low, Claiborne employed fewer people for the Walmart project than it would for its department-store accounts, located them outside of New York City to save on rent, and relied more heavily on technology to keep things "faster, quicker, and cheaper when optimized" at the scale of Walmart's purchases. Even so, Walmart executives persisted in trying to lower the price: "They will ask if that zipper is right, or does that piece of lace trim add value."[28] While Claiborne kept in Walmart's good graces, Rubbermaid didn't, and was punished. In the 1990s, the firm was the leading brand-name maker of items such as kitchen trashcans and laundry baskets. "But when the price for the main component in its products, resin, more than tripled between 1994 and 1996," journalist Leslie Kaufman wrote, "Walmart balked at paying increased prices. When Rubbermaid insisted, Walmart relegated the manufacturer to undesirable shelf space and used its market power to promote . . . Sterilite, which made lower-priced nonresin products."[29] Rubbermaid's profits plummeted and the company was bought by another household goods giant, Newell.[30] Newell managed to get Rubbermaid products back on Walmart's good side.

In the public's eye, Walmart's Darwinian fixation on efficiency in the name of democratically low prices was most apparent in its labor practices. Walmart salaries were quite low, employees were nonunion, and huge numbers of their workers didn't receive health benefits. In regions with a high labor union population, union officials and some government agencies complained that the chain's size and nonunion practices caused benefits and wages in the area to suffer once Walmart moved in.[31] By the late 1990s the retailer had abandoned its "Buy American" mantra and instead obtained most of its products from firms that made them outside the United States at much lower cost. Labor advocates complained that these factories were routinely among those with the worst working conditions. Walmart's CEO in 2000, Lee Scott, adamantly disputed those charges, pointing to a strict code of ethics for overseas suppliers and stating that the company cut off contractors who violated them. "If you are an admirer of capitalism, [Walmart is] the epitome of it," said economist Carl Steidmann of the PricewaterhouseCoopers consulting firm in 2000. "They are the prime example of the good and bad."[32]

Retailing consultants and Wall Street brokers tended to emphasize the good. By the early 1990s the chain's efficiency practices reduced its sales costs to 2 to 3 percent below the industry average—savings that translated into billions of dollars in additional revenue for the company.[33] In 1993 AT Kearney management consultant Burt Flickinger noted that "Wal-Mart is the only car on the track with a jet engine under the hood. . . . [It] is so far ahead it's going to be damn tough for anyone to catch up in this decade."[34] A 1998 article stated that "Wal-Mart's ability to use data to adjust quickly to market conditions and consumer demand has caught the attention of many top business leaders."[35]

Competitor retailers, both large and small, were fearful. Local merchants in small towns where Walmart had pitched its huge tent complained that they could never get manufacturers to sell them goods at near the wholesale prices that Walmart, with its formidable buying leverage, paid. Even the huge Kmart, Sears, J. C. Penney, and Target chains, Walmart's most direct national competitors in the broad discounting realm, struggled mightily to keep up with the market share it had gained at their expense.[36] Department stores were especially vulnerable to competition during the 1990s, when Walmart was spreading more broadly than ever across the United States, and many analysts saw the department-store form as inherently weak. The *New York Times* summarized the view held by many: "Once a growth industry that racked up profits from a rising population tide," it wrote, "retailing has matured into a business where one store's gains now come only at the expense of another. The massive over-building of shopping complexes during the 1970s has left America 'over-malled.' Yet Americans' appetite for these products has slackened. The baby boom generation that turned shopping into a leisure activity is aging, and becoming more interested in saving for the future." Many also thought that traditional department stores, such as Macy's, J. C. Penney, and Burdines, were caught between discounters such as Walmart and specialty stores such as Gap, Ann Taylor, and Crate and Barrel—and even between manufacturers such as Ralph Lauren, Wedgewood, and Burberry that sold their products in department stores. Additional niche competition came from twenty-four-hour ordering departments of catalog companies such as L.L. Bean and Sharper Image. "Department stores are at the end of their life cycle and have to be reinvented," the head of an investment bank told the *New York Times*.[37]

A number of Wall Street takeover specialists exacerbated the situation in the 1980s. Confident they could sell department stores' real estate and unrelated businesses for a sizable profit, this group engineered a succession of department-store buyouts that caused the debt of the new owners to skyrocket, with retail margins that could barely support it. Those huge repayment requirements, noted *Investor's Business Daily*, "hurt [the companies'] ability to upgrade stores and systems to better compete with discounters like Wal-Mart Stores, Inc."[38] The major economic recession in 1990 and 1991 made repayment especially onerous, and the fallout was devastating. From 1990 to 1995 a total of 316 department stores failed, and two such companies, Federated Department Stores and Macy's, entered Chapter 11 bankruptcy protection.[39] A healthier Federated bought Macy's once it emerged from Chapter 11 in 1994. Federated had under its umbrella some of America's most famous regional store brands, from Bullock's in the West to Lazarus in the Midwest to Abraham and Strauss and Bloomingdale's in the East. Many hoped that Federated CEO Alan Questron, an efficiency fanatic, would set an example by steeling his merchants to face Walmart and other discounters.[40]

The supermarket business likewise found it difficult to keep pace with Walmart. A spate of acquisitions during the merger-happy 1980s–early 1990s brought many regional brands under a few prominent names—Kroger, Albertson's, Safeway—while some independent retailers—Harris Teeter, Publix, and Wegmans, for example—held their own as respected forces within the industry.[41] But they all worried about Walmart as its Sam's Club and Supercenters with grocery sections started to open across the United States. The discounter's grocery prices were often set artificially low, as executives anticipated that the store would earn higher margins when grocery customers went on to also shop in the nongrocery aisles. Walmart didn't follow supermarket-industry tradition whereby low margins were made up via a second revenue stream from manufacturers that paid for special placement and promotion of their goods. Grocers historically would charge manufacturers for the right to place certain goods on the shelves ("slotting allowances");[42] take some of the money manufacturers pay for promoting particular products and use it for other purposes ("diverting");[43] and take advantage of promotional pricing by ordering a sale product at a reduced price, then holding it until the sale is over and making higher margins on the regular price ("forward buying").[44] Because manufacturers didn't have to play these sorts of games with Walmart or Sam's Club, they were happy to give Bentonville a better deal.[45] A historical review appearing in 2012 in the trade magazine *Progressive Grocer* noted that the supermarket industry in the 1990s "was put on abrupt notice by the 'Bentonville Behemoth' . . . [t]hat the rules of the game were in the process of being permanently upended by its efficient, and quite literal, march into the food business." That food business, the magazine added, "found [Walmart] at once becoming the most feared, loathed, and respected of any retailer on the planet, before or since.

Indeed, the weakest links in the industry—those that ignored or dismissed Walmart's potential impact on their marketing turf or just refused to evolve to meet new standards—rapidly fell by the wayside in a spate of acquisitions, closures and sell-offs."[46]

Department stores and supermarkets developed a variety of strategies to compete in a Walmart world. Some retailers supported community efforts in their towns to oppose the establishment of the "Bentonville Behemoth." These fights often focused on academic studies and anecdotal evidence concluding that the arrival of a Walmart drove local merchants out of business and devastated downtown areas. Typically these battles pitted citizens who wanted to retain their traditional shopping districts against those who supported the potential economic benefits, such as new construction jobs and the influx of shoppers from outlying areas. Although traditional shopping did win at times, it was an uphill battle. Some local retailers pursued their own programs in the hope of matching Walmart's vaunted efficiency levels: by the early 1990s, the use of barcodes, electronic data interchange, and the so-called just-in-time replenishment of stock was becoming common, especially in groceries.[47] The department-store retailer Nordstrom, for example, touted its rollout in 1992 of an inventory system that used radio-enabled UPC scanning devices and touch-screen personal computers "to help merchandisers make smarter buying decisions and allow them to monitor the unexplained loss of goods."[48] Around the same time, Federated was forging ahead with software aimed at making planning and buying more efficient.[49] These advances could help increase the rate of inventory turnover and reduce the need for markdowns and sales. "That's the area of greatest leverage," agreed Stewart Neill, Saks Fifth Avenue's vice president of management information systems. "If you can speed up turn[over]s . . . that will increase profit more than anything else you can do."[50]

By the late 1990s, according to *Investor's Business Daily,* department stores were turning the corner in efficiency, noting in an article that through the decade they "had been building muscle by consolidating." The article added that "to cut costs and better serve customers they've built clout with vendors in the same way."[51] Nevertheless, when department stores and supermarkets competed toe to toe with Walmart for better margins, they lost. Even its giant competitors Kmart and Sears couldn't match Walmart.[52] These big chains poured millions of dollars into the kinds of satellite systems and back-end technologies Bentonville was using, and yet they still experienced smaller margins, decreasing customer counts, and lower profits. And while Walmart was often first and foremost in the minds of many department-store and supermarket executives during the 1990s and early 2000s, this group also worried about direct and indirect competitors in their own industry. For department stores this meant big discounters and specialty stores, while supermarkets saw Kmart (which carried groceries), limited-inventory discount stores (such as Dollar Store), deep-discount drugstores (such as Drug Emporium), and convenience stores (such as 7–11) as stealing shopper time from their aisles.

In the face of both the efficiency treadmill and considerable competition, supermarket executives began to argue that, beyond the need for back-end productivity, the only way to become profitable again was to find ways to differentiate themselves from the discounters. *Advertising Age* captured the challenge in a 1993 article noting that "much of the supermarket industry seems stuck in the same retailing midway that's creating troubles for old-line retailers like Sears, Roebuck and Co. and local department stores." It quoted an advertising executive who encouraged executives to shift from trying to fill customer desires democratically to following the pragmatic need for discrimination. "No retailer can be all things to all people," he argued. "If you don't have a clearly defined niche in the marketplace, and you don't generate top-of-mind awareness at least among that segment of the market that you say you stand for, you are lost."[53]

Sure enough, retailing discrimination started to catch on. Bill Bishop, founder of the Willard Bishop packaged-goods consultancy, remembers urging skeptical CEO Danny Wegman of Wegmans Food Markets to adopt the barcode system for collecting information on his customers. "When we first started working in the '80s on loyalty data, people like Danny Wegman didn't want to treat people differently," Bishop said. But, Bishop said, Wegman ultimately came around as he began to understand the value of identifying the 20 percent of shoppers who brought in 80 percent of the revenue.[54] The inevitability of discrimination was aided by an existing retail business truism stating that the declining rate of U.S. population growth meant competition for new customers would therefore become substantially more rabid and expensive than in the past. Many in the industry argued that efficiency meant retailers needed to place more emphasis on retaining good customers than on finding new ones—and that the way to do that was to learn as much as possible about those good customers in order to know what persuasive levers would keep them returning.[55]

Most retailing experts also agreed that competing on price alone was not a winning strategy against Bentonville or other discounters. They believed that Walmart was succeeding also because in shoppers' minds its stores were linked to solicitous service. True, pricing had to be low, the experts acknowledged, and to compete successfully certain high-profile items needed to be heavily discounted. But they also cautioned that relying on the lowest price to differentiate one store from its competitors would only encourage a deadly race to the bottom against another retailer that might be better able to sustain losses. Instead, according to a 2003 essay in *Progressive Grocer*, the key was finding a large enough niche in the marketplace that eluded Walmart and that served specific, carefully designated customers and their interests. "Being where Wal-Mart isn't," the article noted, "means targeting customers not served well by Wal-Mart, addressing the needs of that segment in a compelling value proposition, and then designing value innovations that are hard to copy or disadvantageous for Wal-Mart to duplicate. And, of course, the exciting outcome . . . is maximized sales, more loyal customers, and an impenetrable barrier to the encroaching mega-chain."[56]

At its core, this idea wasn't new. Traditionally department stores and supermarkets had attempted to present a particular image to draw certain types of shoppers to them over their competition. In the early 1990s, though, many worried that retailing personas were not sharp enough for the new competitive era. Department stores in particular were singled out: mergers clouded the original brand's personality; store brands were downplayed in favor of nationally advertised merchandise (such as Polo, Etienne Aigner) carried by various merchants; baby boomers didn't share their parents' loyalty to department stores; and department stores tended to follow discounters in focusing on price-oriented rather than image advertising. The lack of a clear-cut image pushed many shoppers into the hands of Walmart, with its constant drumbeat of low prices and service. "Retailers need to prepare to fight harder for their customers," a retailing consultant told *Advertising Age* in 1988.[57] A senior sales executive at Chicago's Marshall Field's department store agreed. "There was a time when we took our existing customers for granted and saw chasing other stores' customers as the key to growth," he said. "Now all that's changed, and we see our future growth coming from our current customer base."[58]

Supermarkets had done at least a little better in establishing individual images. In 1986 a stock analyst told *Advertising Age* that the Albertson's supermarket chain's "niche has been to offer attractive stores, specialty departments, nonfood merchandise, and a little more service at a slightly higher price."[59] With few exceptions, though, supermarket images were not immune to encroachment from Walmart, and stand-alone grocery businesses were losing ground to the discounters that also offered food aisles. This was a new world.

A refurbished loyalty program seemed an obvious solution. But instead of trading stamps, many stores took a cue from airline and credit card companies and offered reward programs, which for retailers meant awarding points based on purchases so that customers could redeem them in the form of discounts and rebates at checkout. "It's hard to walk into a supermarket or department store these days without being asked to join their 'preferred customer' club," noted a 1993 *Advertising Age* article.[60]

Naysayers saw such activities as misguided. A director at the Bain & Company consultant firm considered it a way of "bribing" customers to continue buying, a strategy that he said rarely pays off in the long run.[61] Don Schultz of Northwestern University's Integrated Marketing program agreed. "I see an awful lot of [marketers] who get into it, can't figure out how to make it work and get out pretty quickly," he told *Advertising Age*. Walmart had no loyalty program.

Some urged that loyalty programs be transformed rather than discarded. The real utility in such programs, they insisted, was not in bribing customers but in the valuable insights they could glean about them.[62] Walmart's target audience consisted of shoppers from one socioeconomic background, so competitors that collected data via rewards programs could identify other types of customers and pursue those groups. Shoppers who saved money as a result of their membership in the program would be encouraged to continue buying from the same retailer, while at the same time the retailer would learn much about these repeat customers. Knowing the buying habits of the most frequent and highest spending customers could enable a store to develop its "best customer" profile and therefore its most effective market niche. The same technologies that enabled stores to collect this information also enabled them to reach out to customers with individualized rewards or discounts. And of course other stores wouldn't even be aware of—and so couldn't match—these specific offers. An executive from the Epsilon database company referred to such offers as "a stealth weapon in frequency marketing."[63]

Catalog companies likewise began adopting a targeted approach to reach specific customers. By the late twentieth century these companies faced huge costs associated with producing and mailing their promotional materials. A data analyst for the catalog company Lands' End noted that "we'd go bankrupt quickly" without sophisticated analysis to cull the most likely buyers of items in specific catalogs from the twenty million names in his firm's database.[64] The high cost of producing and mailing catalogs may explain why retailing experts have tended to exhort retailers to find ways to reward the "most valuable (and most profitable)" customers in the store itself.[65]

Department stores were slow to adopt a personalized approach. In the early 1990s they were preoccupied with trying to make their behind-the-scenes operations more efficient as the best way of keeping their books balanced. For them the challenge was to be profitable by offering their target clientele the right merchandise when they wanted it and in the right amounts and sizes, and at the right prices. Still, many department stores did attempt to take what data they had and made targeted mailings based on gender, income, and lifestyle. Taking a page from their forebears, several built personal-shopper programs aimed primarily at corporate customers short on time or fashion sense.[66] Pursuing technologies for in-store personalized deals based on rewards program data didn't gain traction for at least another decade.

In contrast, the idea did begin to take hold in the supermarket business, as those executives were recognizing that individualized shopper attention could prevent defections to competitors. "The average shopper is spending \$50, \$60, \$100 a week in a supermarket and yet the store is not in personal contact with that customer," *Progressive Grocer*'s editor told the *New York Times*

in 1991.[67] Not only did supermarket executives recognize that personalization could establish a level of connection with customers that would keep them from shopping elsewhere, they also believed that their suppliers would pay them to do it. A power shift had occurred in the grocery business, as manufacturers now depended more than ever on stores for reaching people with messages about their products. One reason for the change was the shopping environment; research conducted in the late 1980s and into the 1990s consistently showed that a large percentage—some argued as high as 80 percent—of shoppers made their decisions on specific brand purchases in the supermarket aisles.[68] Manufacturers saw this as an opportunity to use advertising, and especially coupons, more effectively.[69] Procter & Gamble (P&G) CEO A. G. Laffley referred to it as "the first moment of truth," the instance when all the money the company had invested to push its creations was tested.[70]

Traditionally, P&G and other manufacturers of branded products spent greatly on advertising their products in mass media. They also placed discount coupons in circulars that were distributed by mail, in stores, and in newspapers. Americans redeemed roughly 1.7 billion coupons in 1993.[71] Considerable cost was involved to produce and redeem the coupons (each one had to be hand-counted), and the brand manufacturers tended to see them as an unfortunate cultural addiction that they wished they could eliminate.[72] Unable to do away with them outright (though P&G did experiment with this),[73] they tried other tactics to increase their effectiveness. One involved placing UPC codes on coupons for scanning as a way to reduce their processing costs.[74] Companies also began to focus more on increased in-store distribution, which they found could generate four or five times the response rate of newspaper inserts.

Although supermarket operators profited by redeeming shoppers' coupons, they realized they could earn more by allowing vendors to set up advertising or coupon-distribution technologies in their stores and then charging them for a portion of their earnings—typically either a flat fee or a guaranteed percentage (sometimes as much as 25 percent). *Advertising Age* estimated that large chains such as Kroger, which operated 1,234 supermarkets in twenty-four states in 1990, could earn $1 million or more by allowing the dispensing of coupons alone. "This revenue-generating potential doesn't go unnoticed in an industry that operates on a tight profit margin," the trade magazine noted.[75] A horde of technology companies then descended on supermarkets, with various mechanisms for in-store advertising and coupon circulation. Depending on the store and the region, shoppers might see electronic advertising signs, various video information displays, shopping calculators (with ads), ad-supported radio networks, and both shelf-based and checkout-based coupon dispensers. "It's becoming a blur," said an ad agency executive about the profusion of promotional options.[76]

Progressive Grocer estimated in 1991 that 12 percent of the seventeen thousand chain stores and 3 percent of independent grocers offered some sort of electronic coupon program at checkout. Many tried tailoring those coupons to the purchaser's buying habits. Von's supermarket chain of southern California, for example, replaced what *Progressive Grocer* called a "standard" loyalty program in 1990 with one that cross-indexed "by customer and product category" the purchases of the program's three million members. Based on this information, Von's mailed specific promotions to members, matching what they had previously bought with participating marketers' coupons.[77] King's Super Markets sent targeted letters to its reward card holders. "It's possible to do a mailing to kosher customers," said an executive. "It's possible to write to people who never buy pasta to try to win them over." The store even mailed letters to customers who hadn't used their cards in a while in an attempt to rekindle their interest.[78]

By today's standards, these tailored actions seem a fairly simple targeting of customers by segments as opposed to the more current sorts of multivariate analysis that yield very specific insights into shopping habits. Nevertheless, the scanners spewed out so many data points that the industry found itself overwhelmed. The *New York Times* noted in 1991 that "the data is so hard to use because its volume is intimidating and using data of any kind in the supermarket industry is new. The very fact that stores can find out almost anything they wish from the data is bewildering to some managers. Should they pursue lost customers? Senior citizens? Young couples with more than one child? Once they do choose a group, what enticement should they offer?"[79] The manager of Ukrop's Super Markets in Richmond, Virginia, agreed. "You have to be careful because there's so much [scanner data] out there you can be run over by it."[80]

Despite the immense analytical challenges, many supermarket executives believed that they couldn't afford to ignore the personalizing possibilities of database marketing in view of the intense competition among stores. Yet even though the biggest suppliers had the cash and expertise to conduct these database explorations, the grocers typically weren't willing to share their data with them. An exception was a firm called Catalina Marketing, which was enlisted by grocers to connect purchased items to customers only by their loyalty numbers, not by any personal information. Based on Catalina's straightforward analysis of an individual shopper's purchases over time, discount coupons would be printed at checkout using a small Catalina-installed system; manufacturers paid the firm to generate the coupons.[81] So, for example, a shopper who buys shaving cream every six weeks might, five weeks after his last purchase, be presented with a coupon from a shaving-cream brand he didn't use—or from one he did use, depending on the brand that paid for the coupon.[82] In 1994, Catalina was sharing its coupon revenues with seven thousand supermarkets.[83]

Many of the merchants that didn't use Catalina were simply overwhelmed by the amount of data involved or were generally reluctant to adopt personalization. One such grocer, Von's, found it so difficult to conduct sophisticated analysis of its database that the grocer fell back on distributing electronic coupons based just on the presentation of a frequent shopper card.[84] A year later, however, Von's was back again with a rewards program that tried to match shopper purchases with manufacturers' coupons.[85] Because of the consistently low level of statistical capability among grocers, some in the industry believed that supermarkets would ultimately have to bring in sophisticated third parties—even major high-tech firms such as IBM and NCR—to cull purchase information and buy third-party data to encourage an actual "two-way dialogue between the retailer and the customer [in the form of] personalized communication."[86]

By the mid-1990s the level of personalized communication supermarkets conducted with their customers didn't seem that much different from department stores' approaches to their clientele via in-store frequent-shopper point accumulations and targeted mailings. The insulated nature of both industries, their lack of data-analysis capabilities, and the high monetary and competitive costs involved all had stood in the way. But despite the industries' reluctance to embrace the new technologies, they soon concluded that database-driven personalization was a valuable activity that should become an integral part of retailers' relationships with shoppers—if not immediately, then down the road. And as it turned out, developments in electronic commerce were making that road more navigable. Yet the department-store and supermarket industries still needed another fifteen years to make widespread in-store personalization a priority.

The stirrings of electronic commerce date back to the 1970s, and the 1980s saw the creation of a number of key endeavors, including France Telecom's Minitel online ordering system (1982) and CompuServe's Electronic Mall, the first comprehensive electronic commerce service in the United States and Canada (1984). The 1990s ushered in transformative changes that ignited e-commerce along the lines we see today—particularly at mid-decade, when developments reinforced retailers' growing interest in using technologies for personalizing their relationships with shoppers.[87] A signal of things to come was a 1993 *Adweek* article urging the advertising industry to "take note" of the internet. The emerging technology has, the essay said, "the potential to become the next great mass/personal medium."[88] The U.S. government had recently begun to allow private activities on the internet, which had previously been a vehicle mostly for university and military researchers. The major internet-related breakthrough for advertising and commerce was Tim Berners-Lee's 1989 invention of the World Wide Web system of hyperlinks. It led to the creation of graphical Web viewers (browsers), notably (in 1993) Mosaic and, a year later, Netscape, which were the first popular ones. Browsers allowed people to peruse photos and information regarding goods for sale on websites, and to make purchases of those goods directly on the sites.

Mosaic was an apt name for the browser, as websites are typically a mosaic resulting from "a two-way interaction between users and a company's webserver (the computers that host internet content)."[89] When a user types a Web address into a browser, that person's computer is requesting to view the content of a specific page. Although a website's words and images seem to appear almost instantaneously, the conversation between a user's computer and the website's server covers several complicated steps. In general, each item on a page is requested and issued individually. Although the website dynamically generates content, the page itself is "assembled" on the user's computer, which downloads software code from the website and executes it on the user's system in order to create the Web page.[90]

The origin of the content you see on the browser depends on the site that you requested. Sometimes most of the content comes directly from the website you chose—called "first party"—and whose URL appears at the top of the browser. But often at least part of the content is delivered from elsewhere—"third parties," which deliver material to your chosen site typically with the permission of that site. Commercial messages are the most common third-party additions. Of particular interest to advertisers is the Web's ability to identify certain aspects of the person receiving that message even without the person realizing it. When you go to a website this interaction is recorded on the website's server, along with various information regarding the interaction, such as your computer's internet protocol (IP) address—a string of numbers associated with your computer. The IP address can reveal your computer's geographic location within a few hundred feet. The website can also record the browser you are using, the language you are using to communicate, and the speed of your transmission. This information enables the website to tailor commercial content appropriate to your device (that is, the correct language and the right graphical layouts).

But this sort of information, helpful though it is, isn't nearly enough for advertisers hungry for audience information. For one thing, IP addresses aren't useful for identifying a specific user indefinitely. Internet service providers might change home users' addresses every few days, and when multiple computers share one IP address, as is the case with most standard home networks, it's impossible to know whether one computer or several computers are being tracked. One way to be sure of a user's identity was to ask individuals to log in to a website each time, though early on this was not considered a particularly good alternative. Another issue involved tracking purchases on a site. At first browsers were set up so that people wanting to buy more than one item had to buy each item separately; Mosaic had no way to store choices—the electronic shopping cart had yet to be instituted—and so a person could not purchase the items together.

A crucial step in the ability of advertisers to collect information on individuals was the invention of the computer cookie in

1993. Cookies are typically small text files that are placed on a user's hard drive by the website the individual is visiting, and only the person or organization placing the cookie can access it. Even after a user exits out of a website, the cookies placed by the website remain on the individual's computer. Companies quickly realized that cookies could enable them to gather much information about an individual, including the dates and times when the person visits their website. Moreover, if a website could get a shopper to log in or otherwise provide the person's name, email address, or other personal information, this data could also be added to the cookie. The website could retain the information on its servers even if the user erased the cookie (as privacy activists increasingly advocated), enabling it to reconnect the information the next time the person logged in.

Advertising firms also discovered that they could recognize people across sites by placing "third-party" cookies on the user's computer, enabling the marketers to follow individuals from website to website—any site from which the marketers had permission to place cookies—and to record a person's presence and activities on each site. An advertiser, a network of advertisers, or a database company that had made cookie-placing agreements on thousands of websites (or more) could compile a substantial listing regarding websites an individual had visited—and make inferences about the person's lifestyle and interests and how those might affect the individual's purchase patterns.

The cookie was the most crucial of a range of emerging developments that deepened the notion that the Web was a place for promoting products as well as collecting data on individuals and then using that information to entice them to make a purchase.[91]

But despite exhortations from trade journals, retailers approached tailored communication rather irregularly and experimentally, even as the Web quickly became a model for compiling data on consumers. "You don't want your company to be left at the starting gate as more nimble competitors race toward the pot of gold in electronic commerce," lectured a *Computerworld* columnist.[92] An article in *Progressive Grocer* predicted that "everything we currently imagine we know about the consumers will be redefined in the next 10 to 15 years." The article urged readers to "just think of the [small] differences between micromarketing to physical neighborhood and a cyber community. . . . Or consider what service means in a virtual shopping environment in which the retailer never physically meets the consumer, but where they know more about that shopper as an individual than any 'place-based' retailer who sees the consumer three times a week."[93]

But exactly when this complete transformation was going to occur was open to question. The *Computerworld* article, titled "Old Rules Apply in the Marketplace of the Future," cautioned that retailers must create an electronic marketplace "that supports the entire procurement process" and is not just caught up in the enthusiasm of selling as long as the transactions are secure.[94] (Visa, MasterCard, Microsoft, and Netscape were already hard at work on transaction security.) The *Progressive Grocer* envisioned a time frame far beyond the typical five-year business plan. An editorial by the retailing editor of *Computerworld* in 1996 underscored that Web commerce was a long-term wager. "Current betting is that Web-based shopping will be slow to catch on because it is only appropriate for certain types of purchases," he wrote. He suggested that "anything that requires real-time, hands-on examination is probably safe—for now." He predicted that new multimedia development tools, such as Sun's Java, could eventually deliver "Jetsonesque" experiences such as online digital dressing rooms and digital agents to handle "time-consuming" personal shopping. But when these sorts of technologies do become reality, he noted, "like it or not, you'll have to make sure your web presence is second to none."[95]

The efforts of the trade press notwithstanding, many physical retailers were hardly persuaded that the Web deserved their attention, though some were experimenting with an online presence. In anticipation of the holiday shopping season in 1996 some major retailers launched websites, including Sharper Image, Saks Fifth Avenue, F.A.O. Schwartz, Eddie Bauer, J.C. Penney, and Omaha Steaks. Private online services such as America Online and CompuServe also offered electronic stores from companies with physical counterparts, but, according to a *New York Times* article, the overall internet "has a far greater variety." One way a retailer might achieve a Web presence at that time was to join a so-called cybermall, which presented various stores under the umbrella of one master site. In 1996 Saks, Eddie Bauer, and Omaha Steaks were part of a cybermall owned and operated by Time Warner Cable. Such approaches were more prevalent in the pre-Google world, when Web directories such as Yahoo! and search engines such as Infoseek and Excite had yet to optimize their navigation functions. "Finding what you are looking for can be both difficult and time consuming," the *Times* article cautioned, "and many newcomers to the Web will tire of having to jump from site to site and item to item, an experience that has earned Net shopping the reputation of 'Death by 1,000 clicks.' Others will enjoy the adventure and find it like window shopping, without the sore feet."[96] A year later, an article in the *Washington Post* also reflected online retailers' desire to be seen as "growing and quickly becoming more mainstream."[97]

With an increasing number of people using the internet for shopping, the case for physical stores establishing a Web presence seemed apparent. At this time direct-marketing advertisers were setting the pace online, serving ads to individuals based on increasingly specific data pertaining to their online habits as well as their offline habits. (Offline information was purchased from non-Web database firms such as Acxiom.) Meanwhile, online-only retailers such as Amazon were using cookies to show returning shoppers items they might be interested in buying based on their previous activity on the site, as well as on the retailers' data analysis of other people they deemed to have similar shopping inclinations. Some observers began to suspect that online retailers were also using their databases to tailor prices to individuals, though the merchants weren't saying and direct evidence for this was slim. Data collection for commercial purposes increasingly ignited discussions in the press, academia, the federal government, and state governments about privacy and surveillance and their potential limits, and about whether any limits should be voluntary or mandated by law. Web enthusiasts played down any potential harm to consumers and instead touted the benefits of offering ads and products of specific relevance to an individual. They also argued that data-driven Web commerce was a great engine of American economic growth.

Physical retailers observed these developments and recognized that data-driven targeting and personalization held great promise. They increasingly parsed their customer and loyalty data for targeted mailings of sales and catalogs. Yet they were reluctant to spend the tens of millions of dollars it would to take to mount "a web presence second to none," concerned that the Web was actually geared toward catalog rather than brick-and-mortar sellers. Indeed, it was no coincidence that several of the stores featured in the 1996 *New York Times* article had strong mail-order businesses. Firms such as Eddie Bauer and Omaha Steaks were already well positioned for handling online ordering and fulfillment; firms that existed outside the mall, such as Lands' End, were likewise good candidates for the Web. Walmart became involved in 1996 with the help of Microsoft and then revamped the site on its own several years later. Department stores, however, typically were not prepared to compete. Bloomingdale's tried establishing a website, but by 1999 it no longer existed.[98] As of 2004 Lord & Taylor still didn't have one at all. Macy's offered a relatively thin site at first, but in 1999 its parent company, Federated Department Stores, expanded it and also purchased catalog retailer Fingerhut at a cost of $1.7 billion to hone its logistical and fulfillment expertise.[99] A Federated executive noted at the time, "None of us knows how big e-commerce will be. But I'm convinced that the company that gets the most out of commerce will have to have database marketing and order fulfillment capabilities—not just a fancy website."[100]

Although mindful of the direct-marketing model, Macy's executives and their counterparts at other department stores during the early 2000s typically viewed internet commerce as a business to be run parallel to, but separately from, their physical store aisles. The databases for each were typically unconnected. High-end retailer

Bloomingdale's, for example, used a sophisticated system called Klondike to identify and cultivate its top customers who shopped either in the physical store or by calling its toll-free ordering number, but this system was not connected to activities on its website. As early as 1998 Walmart began to allow customers to pick up Web orders in a store near their home.[101] But this policy did not result from enthusiasm for the new retailing form; rather it grew out of a specific need—many customers did not have a credit card and so needed to pay in person with cash. Beginning in 1998 Macy's also allowed store pickup for online orders. In 1998 it brought its website under a new interactive division based in San Francisco. The division's head said that the website had the potential for increasing store sales by attracting younger customers and encouraging purchases from home, especially in markets without Macy's brick-and-mortar stores. Yet the chain's leadership was keen for shoppers to see Macy's.com as separate from the physical establishment so that the former wouldn't be "eating away at its department store trade," reported the *Atlanta Journal and Constitution*. "Macys.com is designed to attract customers who don't like department stores."[102]

In the grocery industry various attempts had been made to sell food online from almost the Web's commercial beginnings. Services such as Webvan and Peapod received much attention for attempting to carve out a niche that involved quick delivery of perishables that had been ordered online. Most lost money (as did many Web ventures at first, including Amazon) and were miniscule in reach. Larger, mainstream grocers such as Kroger and Albertson's experimented with Web-driven home distribution services in the 1990s and early 2000s, but they saw the internet's utility mainly as an extension of their promotion business. In fact, supermarket executives saw the Web as even more distinct from their physical stores than department stores did. A 1998 *Progressive Grocer* article encouraged the personalization possibilities associated with promotional use of the internet. "In on-line shopping models," the article noted, "on-line ads, coupons, new product announcements and a wide range of promotional events can be delivered precisely to specific households with specific demographics and buying histories."[103] Clearly the public was ready to take advantage of ways that the internet could benefit their grocery shopping. The *New York Times* noted that the number of shoppers who visited the coupon-dedicated site Supermarkets.com every month grew from twenty thousand in early 1998 to three million just over a year later. Coupon sites took off when manufacturers realized that shoppers would not go to their corporate websites in search of coupons. They therefore made deals with many sites, including supermarket sites, to circulate the deals. Consumers could print coupons and redeem them at their grocery store. Although traditional newspaper inserts and postal circulars overwhelmingly remained the predominant means for receiving coupons, the *Times* noted that "on-line coupons offer a major advantage: customization. Instead of picking through a random selection of coupons that arrive in the mail or tucked into the daily newspaper, shoppers can fill out on-line questionnaires specifying their interests, ZIP codes and shopping habits and then receive E-mail with relevant offers on a regular basis." Industry analyst James McQuivey predicted that the targeting and personalization of the online coupon business would evolve rapidly. "This is just the first step in the evolution of on-line coupons because there's an enormous opportunity here," he told the *Times*. "In the next two to three years, companies will figure out how to get coupons more and more easily to consumers."[104]

Electronic coupon delivery proved to be a growing part of the armamentarium that supermarket chains used to try and keep customers from straying to other grocery-only competitors or the food-and-everything-else stores of Walmart and other discounters. However, the grocers realized they needed to associate coupon distribution with their own identities as opposed to using freestanding coupon sites. One way to do that would be to present their database-determined valued customers with targeted offers. A variety of emerging firms were engaged to do heavy lifting. The Food Lion chain hired a company called Bigfoot Interactive to provide email communications solutions "for ongoing customer retention and coupon incentive efforts."[105] Foodtown hired S&H Greenpoints, the twenty-first-century incarnation of the popular Sperry & Hutchinson Green Stamps, to help manage its rewards program for points "that can be earned online, offline or via credit card, and can be redeemed at both clicks-and-mortar and bricks-and-mortar stores."[106] And the Great Atlantic & Pacific Tea Co., still in business but far smaller than in its heyday before World War II, analyzed the A&P loyalty program's database and created targeted coupon offers for frequent shoppers. The program, called One-to-One Direct, sent the offers via the U.S. Postal Service.[107]

Inside supermarkets themselves the grocers launched a new wave of targeted-coupon distribution. Prior to 2000, they typically distributed personalized in-store coupons to customers as they exited, most often via Catalina printers. The redemption rate of

Catalina coupons was higher than that of newspaper coupons—6–7 percent as opposed to 3 percent. Still, grocers and manufacturers complained that shoppers lost many of the checkout coupons before they returned to shop again. While not giving up on the Catalina printers some grocers, in response to the growth of general coupon websites, started to target shoppers in the early 2000s as they entered the store or moved through the aisles, mostly through newly installed kiosks. At the kiosks shoppers could swipe their rewards cards in machines that then generated coupons tailored to what the grocer and participating manufacturers believed would interest the customers as they moved through the store. A more sophisticated technology involved radio-connected carts or handheld devices that were linked wirelessly to a store computer and that offered shoppers various supermarket services (such as notifying them when deli orders were ready) as well discounts based on database information; in 2004, both Albertson's and Stop & Shop were experimenting with these systems.

Neither the kiosks nor the other high-tech gizmos caught on. Perhaps they were simply too expensive for stores to justify their use, and maybe customers simply found them too complicated to operate. And possibly grocers were growing a bit impatient with rewards programs overall, which too often were compromised by normal human behaviors. "Quite frankly, we've been hard-pressed to find retailers that are doing the customer loyalty program with specific customer successes," noted a retailing consultant in 2007. "Shoppers forget to bring their cards for every trip, and/or borrow someone else's, and the integrity of the data is then compromised. Pretty soon all the intelligence that an expensive loyalty program has gathered is useless."[108] Still, the logic linking Web promotion to the physical store was clear: if Web marketers could track people's activities and send them coupons online, perhaps grocers could apply the same process even more successfully as people moved through the stores. And for manufacturers anxious to identify likely and valuable customers and to learn as much about them as possible, the ability to personalize a deal as a person faced an actual "first moment of truth" could be an exciting development.

The mobile phone greatly simplified efforts to achieve this goal. For the first time the shopper, rather than the merchant, brought the connecting technology into the store—and that technology could be used to reliably identify the individual. An early approach, the Mobile Rewards system from Boston-based MobileLime, transformed a shopper's cell phone number into a unique identifier. "Employing the cellular-based payment and loyalty capabilities, the single-store operator will offer shoppers instant, individualized rewards without the cost and inconvenience of standard in-store reward cards," noted *Progressive Grocer* in 2005. Mobile Rewards enabled a supermarket "to tailor special promotions and send real-time offers to customers" as they moved through the store; shoppers were sent text messages and were invited to use their cell phone to call a dedicated toll-free number for bargains.[109] In a follow-up piece two years later, the trade magazine concluded that the program was transforming the "lowly mobile phone into a marketing, loyalty, and payment device."[110] In the United States at this time, the plain flip phone (or "feature phone") was by far the most popular type of cell phone among the 77 percent of adults who owned a mobile phone.[111] Smartphones—devices that connected to the internet as well as to cellular phone networks—represented just 4 percent of the mobile market.[112] Those numbers would soon change radically with Apple's release that same year of the iPhone, which reshaped Americans' understanding of mobility and influenced the creation of the open-source Android smartphone operating system that would ultimately dominate the market.

By 2009 only about 29 percent of Americans owned smartphones, but those who did tended to be upscale and were therefore desirable to marketers.[113] Clearly if stores could use the flip phone to identify shoppers and cultivate personalized relationships with them, they could do the same thing with the smartphone, but with far more gusto: these phones could accept cookies from websites, and they could be tracked. At this time most of retailing wasn't yet taking advantage of the smartphone's capabilities, but industry analysts believed that the devices, especially Apple's version, indicated a need to rethink how stores approached shoppers. "Few devices have revolutionized the way people communicate and live more quickly than the Apple iPhone," marveled a *Progressive Grocer* columnist in late 2009. With the advent of home computers and now cell phones, many analysts believed that shoppers for the first time had supreme leverage over the retailer because they could access these technologies to obtain coupons, investigate product quality, compare prices, and make purchases. "The consumer is now in control—think 'Command Center,' noted the *Progressive Grocer.*[114] "The challenge for retailers," the writer emphasized, "is to figure out how to use the technology, and your relationships with shoppers, to the benefit of your [store's] brand."[115]

Supermarkets still considered Walmart to be their main nemesis, but their concern had moderated somewhat when the chain halted its rabid march across the United States in favor of overseas expansion. Some in the industry also pointed out that Bentonville wasn't bulletproof even on a supermarket's home turf. Because the discounter's primary customers were from the lower economic end of the earnings spectrum—albeit a large and growing chunk of America—supermarket chains realized they could pursue more lucrative customers with broader, deeper—and higher margin—arrays of packaged goods. By contrast, they saw Amazon as a growing threat. The online behemoth was a major force in the growing grocery home delivery movement. Amazon Mom, a diaper subscription program launched in 2010, offered 30 percent off diaper supplies, often with free two-day shipping.[116] *Supermarket News* noted that the deal undercut just about everyone, including Walmart. "Thirty percent is a huge incentive to steal a trip and loyalty [from the supermarket] and try to build the mom's [Amazon shopping basket] up," noted an industry consultant.[117]

And it wasn't just diapers, as Amazon began to experiment with home delivery of all sorts of groceries in the area surrounding its Seattle base. Called AmazonFresh, the program offered free delivery for orders of $75 or more (and $5.99 for orders under $75). Further, customers who lived in certain ZIP codes and who shopped once a week and scheduled deliveries on a specific day would receive free delivery with any purchase.[118] The firm had experimented with grocery delivery since 2006, enlarging the service steadily over the subsequent five years. In early 2011 an article in *Supermarket News* reported that the company was mulling expansion of the service beyond the Seattle area. The article quoted an analyst who noted that Amazon could expand its grocery delivery "as far as it wants as long as it has enough of a distribution system, because there's a limit on how far it can go from just one hub."[119] Amazon had, in fact, begun rapidly expanding its warehouses across the country.

Also in 2011 Amazon introduced its price-check app, which grocers realized was a major threat to their control over their store environment. An app that enabled customers to seek lower prices from Amazon as they moved through the aisles of their local supermarket could well result in increased grocery purchases outside the physical store. The widespread concern that permeated the retail department-store industry had now spread to the grocery industry. "Amazon is the biggest thing to hit retailing since Wal-Mart went national," proclaimed the *Journal of Commerce Online*.[120] *Investor's Business Daily* viewed physical retailing's predicament quite ominously. "With e-commerce now a major share of core retail sales and growing rapidly, the writing is on the brick-and-mortar wall," it said. "Adapt or die," the periodical added.[121]

Many believed that the large chains Circuit City electronics and Borders books folded principally because an increasing number of shoppers used these physical stores just to "showroom"—they would go to a store to check prices before making their purchases online from Amazon or other e-tailers.[122] The smartphone (used by about 42 percent of the U.S. population by 2011)[123] was a handy tool in helping consumers compare prices as they walked through the stores. Department-store executives recognized that they needed to make fundamental changes, reported *Investor's Business Daily*. "To survive, traditional retailers such as Macy's [and] Nordstrom . . . are adapting to a fast-changing landscape in which the old one-size-fits-all model no longer works with today's cross channel shoppers."[124] An immediate response was to stop carrying products that shoppers could showroom. The trade press reported that stores began working with manufacturers to modify particular items slightly so that they would each be assigned a unique barcode. Department stores also reverted to their pre–World War II policy of carrying clothes with store labels. Macy's, now the largest U.S. department-store operator, reported in 2012 that 43 percent of all the products in its aisles were either exclusive brands like Madonna Material Girl or in limited distribution and so hard to find elsewhere.[125] Many types of retailers, including supermarkets, could adopt this tactic, but it was not nearly enough; Amazon and other online sellers could come up with similar items that met or exceeded the quality of the physical retailers' house brands, with quick tryout turnaround and at lower prices. *Investor's Business Daily* reported that the house brand was just part of a more comprehensive strategy for competing with the Web and showrooming. Retailers, it noted, had "started to integrate the virtual and physical store worlds to give customers a similar experience."[126]

In the following chapters I will show that this integration is bringing us into a new era of shopping, with the retailing institution tilted firmly toward the ancient merchant impulse of discrimination. The difference is that, in the new world of the universal product code and the internet—and the ballooning amount of personal information that these technologies can generate on virtually any customer—it will be accomplished much more intensely. It will involve an interlocking system of new and old organizations committed to accumulating ever-increasing amounts of data on individual shoppers along with furthering the technologies to track and persuade them. All this information, and all these resources, are changing the industries' understanding of the shoppers who move through their stores, of their stores' layouts, and of the deals they can make with particular customers. The new world of the physical store is linked to a multifaceted approach to buying and selling, and it is challenging shoppers' understanding of the marketplace, themselves, and even the traditional association between shopping and democratic social life that has become such a strong part of American culture.

4 HUNTING THE MOBILE SHOPPER

"Bricks and mortar is dead; long live bricks and mortar!" This was the surprising theme that emerged from the sprawling 2014 Internet Retailer Conference and Exhibition, reported Laura Heller, editor in chief of the online trade magazine *FierceRetail*.[1] Everybody in retail had previously been emphasizing only the tremendous advantage that digital retailers had over physical stores because of the vast amounts of customer information they could amass and exploit. True, brick-and-mortar stores also could size up customers by recording their purchase history and by purchasing additional information about them from data brokers. But digital sellers could do the same—and a whole lot more. They could follow customers around the internet and track them at any site where they had shopped, and they could also present them with ads on other sites for the same items the shoppers had viewed earlier on the merchants' site. When those shoppers returned to the sellers' website the merchant could offer up the products they had previously considered and possibly rejigger the deal—change the price, add other items, reconfigure the warranty, and more—based on the shopper's assessed value to the business, such as whether the person was a consistent customer yielding good margins for the store, a bargain hunter, or someone in between.

Given the seemingly insurmountable edge enjoyed by online retailers—beyond their extraordinary and ever-growing database of information on individual shoppers, they also enjoyed virtually unlimited shelf space—some wondered whether physical retail real estate was dead weight, a hindrance to merchants attempting to compete with seemingly ever-flexible digital companies. But Heller noted that the zeitgeist was beginning to change. The attendees at the internet retailer conference, she said, were recognizing the physical store as "increasingly a critical part of an online or omnichannel strategy" (*omnichannel* is an industry buzzword that refers to the multifaceted means by which a physical retailer can profitably meld its online and mobile phone capabilities with its physical resources).

The largest department stores, supermarkets, and discounters were well aware that online-only stores were continuing to move in on their turf. The success of Amazon was probably the biggest factor behind physical merchants' new focus on beating internet rivals at their own game. Succeeding at this gambit was not going to be easy, however. At the end of 2014, Amazon announced that more than ten times as many items were ordered with same-day delivery for Christmas than in the previous year. Moreover, the company reported that nearly 60 percent of its customers had purchased from Amazon using a mobile device and that the percentage of people using mobile shopping increased as Christmas drew closer. Total U.S. holiday sales from the Amazon app doubled in 2014.[2]

Major physical retailers understood that developing winning websites and apps was necessary to countering this challenge, but they saw it as just one part of a larger overall strategy. Believing that a material store in one form or another would always be a key to success, retailing executives wanted to encourage shoppers to buy items from their websites or apps and then pick them up from their brick-and-mortar stores. More important, they decided to copy what internet retailers had been doing: tracking shoppers to discern their lifestyles and the goods that appeal to them; distinguishing individual shopper value based on that information and the shopper's prior purchases; and (depending on a shopper's assessed value) offering targeted enticements, which could include altered pricing, to encourage both a relationship and a sale.

As late as 2013, many brick-and-mortar chains continued to back away from direct confrontation with internet-only companies. Executives, especially those overseeing major department-store and discount operations, instead tried to work around online sellers by offering merchandise that their shoppers would not be able to find anywhere else. For example, they increased their stock of exclusive brands or had their suppliers tweak the products they produced for them just enough so the items would be assigned new barcodes.

After 2013, these companies shifted their focus and began to build parts of the business they felt could beat or at least match online-only merchants. For one thing, they designated part of the physical store as a warehouse. The idea was not new; during the 1990s and early 2000s Walmart, Sears, and Macy's were among retailers that had instituted systems enabling shoppers to buy online and pick up at a nearby chain location. They called the practice "shop online and ship to store," or "click and collect." But many other merchants demurred. The chief executive of supermarket giant Kroger told investors as late as October 2013 that people went online for recipes and coupons and to check their loyalty reward points, but "not for e-commerce." The international retailer Ahold, whose U.S. chains include Stop & Shop and Giant-Landover, had a different view: it had already instituted its Peapod service, which enabled customers in select areas to order online and pick up the goods at one of its nearby supermarkets.[3] Nevertheless, as far as Kroger was concerned, "E-commerce makes up a tiny, tiny percentage of the food business."[4]

By 2015, though, the online purchasing and in-store pickup system was becoming commonplace for most sellers with a national footprint. Even Kroger was now moving in that direction with an experiment in Houston.[5] Department stores and large discount chains were also touting their ability to accept returns from internet purchases at the physical store. Astonishingly, in 2013 customers returned an estimated third of all purchases they made on the internet,[6] and this proved to be an opportunity for physical stores to shine: surveys indicated that customers liked the choice of returning goods to any branch, regardless of where they had made the purchase.[7]

With these new developments physical retailers started going head to head against their biggest internet-only competitors, who wanted to make brick-and-mortar stores look anachronistic. Amazon, eBay, and Google Express delivery service instituted a new arrangement by which they contracted space in shopping centers, stores, and bus stations for pickups and returns, but the program did not initially catch on with shoppers.[8] Instead, a number of online-only businesses—grocery delivery service Instacart, Google Express, and eBay, among others—focused on beating brick-and-mortar's alleged location advantage by targeting delivery to home or office.[9] They aimed to make their services so convenient that shoppers would not even have to go to a physical store to purchase the most basic goods.

Amazon was at the forefront of such efforts. In 2013 the company announced a plan to use drone aircraft to deliver packages to customers' porches, lawns, or roofs within thirty minutes of an order being placed. Although the U.S. Federal Aviation Administration initially balked at the notion of drones flying around with parcels, it eventually allowed these and other unmanned initiatives in airspace below four hundred feet.[10] In a more conventional move, the company began offering free two-day shipping in 2005 on over twenty million items via its Amazon Prime program (memberships cost $99 a year in 2015). In 2009 it began offering same-day delivery in Manhattan, Philadelphia, San Francisco, and a few other heavily populated areas. And for the 2014 Christmas holiday season the company offered free two-hour and $7.99 one-hour delivery service in Manhattan for one day between the hours of 6 a.m. and midnight. Twenty-five thousand products were eligible for this service, including paper towels, shampoo, books, toys, batteries, USB cables, and Kindle Fire tablets. Bike messengers made many of the deliveries. Amazon planned to expand the offering to other large metropolitan areas.[11] By offering same-day delivery the company was hoping its customers would decide to "skip the trip to a retail store," an executive at a mobile apps production firm told *Mobile Commerce Daily.*[12]

Dave Clark, Amazon's senior vice president of worldwide operations, confirmed this goal: "There are times when you can't make it to the store and other times when you simply don't want to go. There are so many reasons to skip the trip and now Prime members in Manhattan can get the items they need delivered in an hour or less."[13]

The brick-and-mortar side fought back. Walmart,[14] Macy's, hhgregg, and others moved to match the quick-delivery programs offered by Amazon (and others). Mall operators such as Simon and Westfield also stepped in to help their smaller tenants compete.[15] The move to quick delivery involved the implementation of new processes and technologies on a local level, and it also meant that part of the store had to be turned into a kind of mini-warehouse. By 2015 stores were still mostly testing their programs in small patches of the country, but the two sides seemed to achieve a rough parity in the ability to, literally, deliver the goods. Analysts noted that while these efforts were expensive and might not even distinguish any one retailer over another, same-day delivery by physical retailers was probably necessary. "Like other traditional brick-and-mortar retailers, Macy's is adapting to a digital age in which people are growing accustomed to getting what they want when they want it," reported CBS's "MoneyWatch." "The chain probably has little choice but to face this new reality and take a risk on same-day delivery."[16] An executive at the grocery consultancy Willard Bishop noted the same situation with supermarkets. "It's messy, but it's inevitable," he said in 2016. "Stores that do it well are reporting incremental gains of up to 30 percent when shoppers use them for both physical and online shopping. But it's still not profitable."[17]

As the battle continued to rage, both sides ignored a crucial terrain: the selling floor. It was there, amid store aisles, that about 90 percent of all retail goods continued to be bought and sold as of May 2015, according to the U.S. Census Bureau.[18] And just as some brick-and-mortar retailers nervously turned part of their stores into warehouses for the internet world, they viewed their aisles equally apprehensively as domains that, they felt, would also have to function somehow like the internet. Although the notion that a store aisle could serve in such a capacity might seem odd, merchants had concluded that this would be the new role of the physical store in the digital shopping era. "Savvy retailers are using multimedia content to make the shopping experience more engaging," wrote Justin Honaman, a partner in a company that provides customer analysis technologies, in January 2014. "They will offer in-store only events and services . . . aimed at bringing customers to stores and keeping them there."[19] He also predicted that personalization "lies at the center of omnichannel marketing strategies in 2014," pointing out that retailers now had the means to "reach shoppers with what they want, when they want it and via the channel they prefer to shop. Retailers will also work to curate the right product information at the right time in the shopping process."[20]

To achieve these goals, he wrote, retailers will need to "connect the dots . . . and [create] a comprehensive customer profile." According to Honaman, retailers with physical stores actually had a unique advantage over their internet-only competitors. While each group could compile mountains of data and build individual profiles by tracking shopper behaviors and purchases online, and then target personalized ads and discounts, only physical retailers could exploit this gold mine of customer knowledge and engagement when customers went from the virtual to the material world, as they so often do. He argued that brick-and-mortar emporia had to be seen as an integral part of a new retailing mandate to "reach shoppers with what they want, when they want it, and via the channel they prefer to shop." For example, a physical retailer could offer an individual the most effective prices "both in-store and online," and in a real-time, personalized manner.[21]

In Honaman's view the key to success was the cellular phone. "Consumers are increasingly using mobile phones and tablets for product research and online purchase," he wrote. "The combination of portability, connectivity and relative affordability gives the smartphone a privileged place in driving always-on commerce."[22] Indeed, the smartphone was already becoming part of retailing strategy. "Mobile is the new first screen," noted a *MediaPost* article at the end of 2014. "On average, people check their phones 150 times per day. More than one third of customers are 'mobile only.' With stats like this coming out daily, we (should all) get it." That meant, the writer said, developing mobile strategies for engaging with the shopper for "loyalty and revenue."[23]

Retail establishments have long audited shopper movements in their stores. The turnstile was perhaps the most common mechanism used to measure the activity at a store's periphery during the twentieth century. The first use of turnstiles in contemporary shopping has been traced to Clarence Saunders, who installed one at the entrance to his first Piggly Wiggly store in Memphis, Tennessee, in 1916.[24] Turnstiles ultimately fell out of favor in retail venues, and by the end of the century some merchants were using electronic beams to tally the number of people who entered their establishments. In addition some stores, especially supermarkets, worked with product manufacturers to use a light source to count the number of people passing through particular aisles, or areas within aisles. (In retailing the familiar term "foot traffic" has now morphed into the rather scientific-sounding "shopper analytics.")

Retailers accepted that this form of tracking wouldn't include identification of individual customers. In addition, the ten largest companies that engaged in this form of tracking agreed in 2013 to a "Mobile Tracking Code of Conduct" that required stores to post conspicuous signs when using tracking technology, and to offer a website from which customers could opt out of being tracked. Plus, data collected could not be personal or "be used to adversely affect a shopper's employment, healthcare or insurance."[25] But with ever-evolving tracking technologies and intense competition among these companies, the veil of anonymity began to fall away.

Consider the experiences of two such companies, ShopperTrak and Euclid. ShopperTrak, founded in Chicago in 1991, devised systems that solved two major problems: the inability to distinguish both *things* (such as baby carriages and shopping carts) from people, and individual shoppers entering a store side by side or en masse.[26] By 2013 the company had developed a small video camera with microprocessors that not only could detect individuals entering side by side or in groups, but could also distinguish between children and adults. In interviews for this book ShopperTrak executives said they were careful to build anonymity into the technology—the system did not capture faces—because they believed their clients were wary about intruding on shoppers' privacy.

Euclid Analytics, a 2009 Silicon Valley startup, came up with an entirely different approach. An early force in accessing shoppers' smartphones, Euclid could follow individuals' movements throughout a store using Wi-Fi, which at the time was in use by more than 90 percent of smartphones. (Wi-Fi allows a mobile device to connect to the internet via a wireless network access point—a "hot spot.") Smartphone owners typically keep their phone's Wi-Fi access turned on, often so they can use the free Wi-Fi available at many retail establishments. And indeed, retailers have offered this service at least partly because they know it makes the phones identifiable. Smartphones constantly ping in search of Wi-Fi network access nodes, and these pings contain the smartphone's MAC (media access control) address—a unique string of letters and numbers that identify the device. Euclid developed a system that could be installed in its clients' stores to follow pings, enabling the company to note the presence as well as the location of every smartphone in the store. (The smartphone owner remained anonymous, however, because Euclid scrambled the MAC number.) The tracking also extended to exterior display windows so that retailers could attempt to monitor the effectiveness of displays in attracting customers into the store. Euclid could also provide the total amount of time a customer spent inside the store—helpful information, as studies have found a direct correlation between the amount of time shoppers spend in a store and the amount they spend. In addition, Euclid contended that its data could help floor managers allocate personnel based on where and when the number of customers would be greatest.[27]

Bill Martin, one of ShopperTrak's founders, didn't believe that tracking the location of individuals as they moved through a store would result in greater profits for the merchant, given the cost of the technology. However, many of his clients nevertheless wanted ShopperTrak to extend its tracking beyond the store perimeter.[28] In 2014 the company responded by buying RapidBlue Solutions, a Helsinki firm whose technology was similar to Euclid's. By combining their systems and adding new features, ShopperTrak hoped to generate a larger sample of individuals than Euclid was getting by Wi-Fi alone.

The company's new approach involved identifying and following mobile devices via any of three signals smartphone owners might have activated on their handset—Wi-Fi, Bluetooth, or cell tower connections. Like Wi-Fi, Bluetooth is a wireless technology with a consistent ping, though it has a less powerful signal and so is typically used for exchanging data over shorter distances (say, between a cell phone and a nearby music speaker). Still, the company believed that Bluetooth could be useful for tracking devices near or in a store when Wi-Fi is not turned on. If neither signal was active, the company could still track shoppers using the phone signal, which bounces off the nearest cell tower, and which is also broadcast with the unique MAC number. At best a phone signal can establish location within 150 feet, while Wi-Fi offers accuracy as precise as 10 feet, and Bluetooth within 10 feet. A ShopperTrak executive said the company could narrow down a cell signal enough to detect a phone in front of or near a specific retail location.[29] He acknowledged, though, that cell-signal tracking would not work well inside a store.[30]

The company's technologies faced various barriers set up by phone manufacturers. In the case of following phones via

Bluetooth, the manufacturers often assign random and rotating false MAC addresses to disguise the device's identity so it can't be identified by a Bluetooth transmitter. Only when the phone's owner, or an app the owner has loaded on the phone, reaches out to the transmitter does special software allow the devices to convert the random MAC addresses into the real one that allows the two devices to communicate.[31] Tracking people via Wi-Fi also raised issues. Apple voiced concern that some marketers were violating iPhone owners' privacy by pinging Wi-Fi signals of iPhones to get MAC addresses even though they had no apps to link up with those phones. Instead, Apple alleged, the marketers wanted to collect MAC addresses of phones and then find ways to link the phone IDs with personally identifiable information—names, email addresses, postal addresses, and the like—and use the data for directly targeting individuals.

Also around this time, many citizens were taking notice of—and objecting to—this form of tracking. In 2013 customers and privacy advocates complained after Nordstrom posted signs at seventeen of its stores regarding tracking that Euclid was conducting at those locations. The retailer tried to reassure the public that the data gathering was benign, stating that "anonymous information" giving Nordstrom "a better sense of customer foot traffic" would ultimately lead to an enhanced shopping experience. Ultimately Nordstrom declared that it was ending its "test."[32] Euclid stated on its website that public concerns about its activities were unfounded. "Our sensors collect only basic device information that is broadcast by Wi-Fi enabled phones. This does not include any sensitive data such as who you are, whom you call, or the websites you visit."[33]

Not everyone agreed. One critic noted that, even if those firms were not intercepting individuals' data, leaving a phone's Wi-Fi turned on "can let your phone leak all sorts of useful things for malicious actors to intercept and act upon."[34] U.K. technology writer Julian Bhardwaj previously wrote that "it's very likely that your [Wi-Fi enabled] smartphone is broadcasting the names (SSIDs) of your favourite networks for anyone to see. This alone might be enough for someone to glean information about you: where you work, where you live or your favourite coffee shop for instance."[35] Even worse, an attacker might use software to kick you off a legitimate Wi-Fi connection and then, when you try to reconnect, force your device onto a rogue network that, on the surface, seems very much like the legitimate one. The goal is a "man-in-the-middle" attack: intercepting data sent between you and a friend, or your bank, and "giving the impression you're talking to each other over a private connection, when in fact the entire conversation is controlled by the attacker." That situation could cause major damage, such as stealing your personal data (for example, bank account or credit card information) or installing malware on your cellphone. "Whilst we have all embraced the technological age none of us are really experts in the field and therefore we are extremely vulnerable to attack," noted an expert on issues related to mobile environments linked to Wi-Fi networks.[36] Bhardwaj noted that "there doesn't appear to be an easy way to disable active wireless scanning on smartphones like Androids and iPhones."[37]

Apple attempted to address this situation in September 2014 as part of an update to iPhone's operating system (iOS 8). The system now transmitted two MAC addresses during Wi-Fi pings—of which just one was correct. Apps on the phone searching for the Wi-Fi link would latch onto the correct code, but a company simply trying to exploit the phone for other purposes would be fooled.[38] Critics complained that the presentation of false and true codes was too patterned, and that with some phone settings the "spoofing" didn't even work. Nevertheless, the update threatened to derail ShopperTrak and, especially, Euclid. A Euclid representative downplayed the development, which she said involved just a subset of the iPhone universe. She also pointed out that Google's popular Android smartphone operating system did not have that Wi-Fi restriction. Consequently, she said, Euclid had thus far not lost much valuable data about shoppers' movements.[39] A ShopperTrak executive similarly played down phone manufacturers' attempts to stop firms from tracking phones through Bluetooth. Not all phones had Bluetooth software that used rotating MAC addresses, he said, so ShopperTrak could still follow that subset through the store.

Nevertheless, Apple's potential destruction of marketing-tracking vehicles was a sober reminder to both Euclid and ShopperTrak that technologies in the digital space are fluid and dependent on the whims of those outside the trackers' control. Some believed that Apple had made the change because it was irked that marketers were finding ways to use its operating system without passing some of the revenue along to Apple.[40] Others believed that Apple wanted to move away from Wi-Fi and original Bluetooth and toward a different version of Bluetooth technology, Bluetooth Low Energy (BLE). BLE uses considerably less power and cost than original Bluetooth, while maintaining a similar range. Companies can buy inexpensive BLE boxes, which act as beacons, transmitting a signal with a device ID. If a phone app within that range is compatible with that ID, the signal alerts the app to send a message via cellular or Wi-Fi that the phone has made a connection with the BLE beacon in a particular location. With an array of its BLE beacons tuned to its app in a retail location, the app owner can therefore figure out the movement of the phone's holder as she or he moves through the store.

BLE can function with many operating systems. Apple engineered a version for apps that were available from its iPhone stores as well as one for the iPhone's built-in Wallet app (originally called Passbook). The Wallet could hold discount coupons and other messages related to trade and travel from retailers and manufacturers. The marketer could attach a location tag to the message so that a reminder about the message would appear on the phone's lock screen (that is, the screen that users first encounter each time they want to access their phone; they first have to swipe a finger across the screen or engage in some other action in order to unlock it) when the phone entered into the range of the BLE installed in the store. Marketers especially liked the lock screen feature because Apple allowed the message to be displayed so long as the tagged message was in the Wallet—the device didn't need to actually carry their app. They therefore saw their challenge as encouraging people to load coupons and other messages into Apple Wallet.

Both ShopperTrak and Euclid began to consider BLE as the best option for tracking foot traffic in and around stores, as neither Apple nor Google were likely to derail it. There was an obvious challenge, however: Apple had built the technology so that only the firm controlling both the BLE transmitter and the app could identify the phone, and neither ShopperTrak nor Euclid would likely be able to persuade smartphone owners to install the app of a foot-traffic auditing firm on their phones. To get around this issue, a high-level ShopperTrak executive said in 2015 his company was considering a plan to install its own Bluetooth beacons in stores while arranging to incorporate its location software tool into the apps of unrelated companies' phone apps. As the foot-traffic auditor of 55,000 retail establishments around the United States, the executive seemed confident that both stores and app companies would be agreeable: when its app is pinged by a ShopperTrak beacon, the app owner will learn the exact location of one of its shoppers, and the company can then deliver an ad on the spot to the smartphone as well as add new information about the phone owner's shopping habits to its data files. Privacy, the executive argued, was not an issue. People who download apps and turn on the Bluetooth function should assume they might broadcast their presence via apps on their phones, he reasoned. He emphasized that ShopperTrak itself would still not be collecting identifying data or retaining any of the data in ways that would link back to the device. Foot-traffic monitoring, he said, had to be above suspicion to people moving through stores.

The actual situation was not so straightforward. As with Wi-Fi, some experts worried Bluetooth technology could likewise make smartphones vulnerable to anonymous third parties with man-in-the-middle intentions. The fear was that a malicious app on a person's phone could sniff out the unique ID of a retailer's beacon and then act as if it was the correct app to connect the phone with the beacon. "The user will never know that there was an app that was looking at where they were, capturing analytics and pushing them very targeted advertisements based on what they've seen," worried the CEO of a company that has installed beacons in many retail establishments, airports, and hospitals.[41] A number of Bluetooth providers say they have since solved this problem.[42] Yet ShopperTrak's plan included a different means for identifying and tracking an individual without the person's knowledge—making this information available if not to ShopperTrak, then to the company whose app ShopperTrak was using. If, for example, ShopperTrak made a deal to incorporate its beacon software into the app of a social media firm, that social network would then possess unprecedented real-time knowledge of members' precise locations.

Not surprisingly, traffic-analytics firms quickly recognized that tracking people for auditing could also mean tracking them for marketers. Raul Verano, chief technology officer for the analytics firm Shopperception, said that his company had installed cameras with 3D sensors in stores to track shopper activity in proximity to goods made by the company's clients. In deference to U.S. and European privacy concerns he said that the company did not analyze the images for age and gender. (The policy was different in Asia, where privacy concerns were low, he said.) He also emphasized that the company discarded the U.S. and European images at the end of each day. "You're so anonymous with our system," he said, "that when you go away from the detection area to another camera area, you can't tell it's the same person."[43] He said that the firm had been experimenting with BLE, sending ads to shoppers on their phones based on data picked up by the camera sensors, and the company was planning to piggyback onto other companies' apps to attract increasing numbers of clients. Like the ShopperTrak executive, he likewise contended that because shoppers had their Bluetooth function activated and were using their apps while out shopping, privacy was not an issue. Although the company could follow people and send ads to them without having particular data about them, he said that the trend was leading toward personalized messages. He predicted that business-as-usual for companies like his would ultimately include both an analytic component as well as personalized advertising.

While ShopperTrak, Euclid, and even Shopperception worried about capturing personal data via foot traffic, another company, shopkick, had no such qualms. It was one of a new group of technology-inspired firms that dived zestfully into the new playground of mobile customer profiles. The tale of Cyriac Roeding's founding of shopkick in 2008 underscores how quickly retailers embraced him for his new take on loyalty. A German-born student of engineering and business who had been impressed by mobile phones as "the next big thing" while in Japan during the mid-1990s, Roeding decided to look for "an idea that has the potential to become a really large company in mobile." He ran CBS Mobile (CBS Corporation's wireless business for its entertainment, sports, and news concerns) for a while in the 2000s, but shifted to the position of entrepreneur-in-residence in 2008 with Silicon Valley venture capital firm Kleiner Perkins Caufield & Byers. It was there, Roeding says, that he saw a gold mine in mobile tracking as the solution to the central problem of retailing. "The number-one challenge facing every retailer in America is getting people through the door," he told *Entrepreneur* magazine in 2011, half a year after shopkick launched. The next critical step, he realized, is "conversion," the industry term for a shopper turning into a buyer while in the selling space. "Conversion rates in the physical world are so much better than online—between 0.5 percent to 3 percent in the virtual world and between 20 percent and 95 percent in the real world. So if foot traffic is so important, then why hasn't anyone rewarded people for visiting stores? The answer is simple: It's because nobody knows you came through the door."[44]

Roeding's comments gave no indication that he was aware of foot-traffic auditing—or perhaps he simply didn't want to acknowledge the earlier model and its adherence to anonymity. He represented himself as part of a new business and approached retailers, brand manufacturers, and investors with a four-pronged plan:

- Award points ("kickbucks") to smartphone owners with a shopkick app for going into a store, for handling products in the store, and for buying those items;
- Use the shopkick app to notify the company every time a specific smartphone crosses the threshold of the store;
- Reward a shopper with kickbucks for scanning products in the store and for purchasing products. Charge the store a small fee for each kickbuck a shopper earns from scanning. If the shopper makes a purchase after using the app (eventually, the shopper could make purchases *through* the app), charge the store a small percentage of that transaction;
- Gather as much data about app users as possible, sharing some of it with the retailers as an added incentive, but use the information mainly as a data pool to help shopkick present highly refined profiles of individuals to brands and retailers wanting to advertise to them in applicable shopping locations via the app.

The shopkick app that Roeding's company created prompted users to register the items they handled in stores by scanning the items' barcodes. The notification signal to be sent to shopkick by smartphones entering the store posed a technical challenge, because at the time Wi-Fi signals could invariably leak outside a store and be picked up by passersby. shopkick engineers solved the problem by developing a device that was mounted near the store entrance and emitted a signal undetectable to the human ear but was picked up by a smartphone's internal microphone—and recorded by the shopkick app. Because the signal was limited to the store's perimeter, the store could be sure that the kick-bucks earned for entering were legitimate. The third part of the plan—the fee shopkick charged the stores—brought the Green Stamps loyalty playbook into the twenty-first century: now merchants could trace these electronic stickers directly to an individual customer's buying record. Retailing executives were drawn to shopkick because it both collected data about individual shoppers and distributed targeted advertising close to the product—proximity marketing, as it became known.[45]

shopkick launched in 2010, with Macy's, Best Buy, American Eagle Outfitters, and the large Simon mall company trying out the service in some of their locations. The actual dollar amount in terms of rewards that shoppers could receive was typically quite low—reports suggested that people on average gained $3 for their assiduous work over many store visits—yet the app's popularity quickly grew, surpassing eight million users by 2014.[46] During this time shopkick expanded its business model by adding BLE boxes to its tracking system so that it could present shoppers with offers that would be delivered via their smartphones as they approached targeted products. In a test on shopper behavior at American Eagle Outfitters, some shopkick users were notified upon entering the store that the merchant was offering "a small incentive" for visiting the fitting rooms. The company reported that more than twice as many shopkick users who received the message visited the fitting room area as those who didn't.[47]

Merchants were especially drawn to shopkick because, unlike other apps relying on BLE, users didn't have to open the app for it to operate. Newer features on iPhones and some Android devices enabled shopkick's engineers to make their app "wake up" automatically when in the range of a BLE and get right to work sending personalized messages to the user.[48] shopkick also encouraged its users to indicate their interest in specific products while browsing all the items that can be viewed on the shopkick app, which would then remind them when they entered a store that sold one of those products.[49] A surprise discount on the app could then catalyze a purchase.

Competitors that were determined to follow shopkick's model and convince millions of people to download a marketing app found the challenge difficult. inMarket succeeded by instituting an aggressive but very different strategy. Unlike shopkick, it didn't expect to reach shoppers just by building loyalty into its own app. Instead, it made deals with many retailers to place BLE boxes with the inMarket code throughout their stores. The deals also included the right to put inMarket code in those retailers' apps so when shoppers placed the apps on their phones they could be pinged by the inMarket boxes as they moved through the stores. Extending its reach, inMarket also arranged for the owners of other popular apps—for example, the magazine publisher Conde Nast and the newspaper publisher Gannett—to include inMarket code in their apps. Their incentive for doing that was to share shopper data and money. Having BLE boxes with inMarket code in thousands of places that could connect to millions of phones with its code meant that inMarket could trace millions of people's movements through stores and send advertising messages to them as they walked around the stores. Coca-Cola and Hillshire Brands were clients.[50] The retailers would get a slice of the advertising revenues if an ad appeared on a smartphone via their app or while a shopper was in their store. inMarket divvied up payments from advertisers to the firm that owned the opened app with the ad, the company that sold the ad (typically, inMarket or the retailer on the retailer's app), the retailer where the app opened, and inMarket. Retailers would be able to set the order of inMarket-linked apps to be woken by the beacon. So, for example, a person walking into Marsh supermarkets with Bluetooth on would trigger the Marsh app. If the phone did not have a Marsh app, the WebMD app might wake up, if it was Marsh's second choice and it was installed on the phone. If Marsh's retail advertising sales force had sold a Nyquil cold medicine ad for the beacon near the pharmacy, inMarket would send that ad to the phone via the shopper's WebMD app.

Apart from advertising revenues, inMarket's website promised to deliver to retailers "rich insights into shopper behavior"—for example, age, gender, visits by hour and day of week, "and more."[51] There was a lot more, as the company collected a lode of data from shoppers in different ways.[52] For example, if a Nyquil ad were served in one of inMarket's own apps (FreeCoupons and ListEase, among others) inMarket would get full personal information about the individual. It would also get the ability to add to the shopper's data file information such as when the person entered the store and the nature of the ad viewed. If, however, the person used the WebMD app, only WebMD could collect the personal information, as stated in its privacy policy. Although that person would be anonymous to inMarket and Marsh, the companies would still get potentially useful data. They would learn the app ID related to the device, which would allow them to record the presence of the device wherever the person was situated in store (in the case of Marsh) or (in inMarket's case) wherever the firm placed beacons. They would also know the person's location in the store, and the ad the person saw. They could store that material, and create a profile based on the history of the person's anonymous movement in the store. And if the person with the phone happened to fill out a sweepstakes form on the phone's WebMD, Conde Nast, or Gannet apps—which would, of course, include the individual's name and address—that person's anonymity would dissolve, thereby enabling inMarket to purchase personal information about the shopper from third parties.

By late 2014 inMarket claimed to have 31.5 million monthly users. In comparison, Walmart was reaching twenty million phones a month with its app, and Target, Walgreens, and Kroger each had approximately four million users of their respective apps.[53] Substantially fewer people were loading other retailers' apps, with the result that those stores could not take full advantage of the tracking and personalization technology. Lord & Taylor was one chain that decided to contract with a number of cross-retailer apps deemed fitting for the store's image, rather than create a stand-alone app. Indeed, inMarket, SnipSnap, Swirl, and other beacon-oriented marketing firms saw fertile terrain in small- and medium-sized businesses lacking the scale to go it on their own.[54]

Giant retailers had an advantage, as they already had widely circulated apps before installing BLE beacons. They also had the means to compile an arsenal of data to profile their customers' shopping habits, so much so that sometimes they could even infer how shoppers moved through the aisles. These retailers typically collected personal information based on loyalty card (or app) registration or via credit (or debit) card identification customers provided at checkout. They captured a record of the items an individual purchased by linking the UPC codes of the purchased items to the purchaser. They then linked the online and mobile activities of every shopper—their purchases as well as what they viewed and/or commented on at the retailer's website and possibly elsewhere—to purchases they made in the physical store. Walgreens' privacy policy acknowledged outright that the company was collecting personally identifiable information by connecting purchases to app and loyalty-program registrations (including information obtained via customers' Walgreens-purchased health devices that linked to the company's app).[55] Walmart implemented cutting-edge customer analytics to understand how online behavior influenced in-store behavior, and vice versa.[56] Target also used sophisticated analytics, collecting and purchasing personal information about individual shoppers from just about any available source—"including information you submit in a public forum (e.g., a blog, chat room or social network)." It also reserved the right to share the information "with other companies . . . to provide special offers and opportunities to you," if a person didn't opt out.[57] A *New York Times* article suggested that data Target had gathered on a young unattached woman based on her recent purchases enabled the retailer to infer correctly that she was pregnant, and proceeded to send her coupons related to her condition—which her father intercepted.[58]

The supermarket industry was moving forward with a range of apps and vigorous data collection even before beacons exploded in popularity. Retail executives widely praised Kroger, the nation's second largest grocery company (behind Walmart), for its ability to compile information about its customers. To accomplish this it enlisted dunnhumbyUSA, a highly regarded analytics firm it co-owned with the British company Tesco. (Kroger ultimately bought out the U.S. branch in 2015 and renamed it 84.51 Degrees to reflect both the quantitative nature of its business and the coordinates of Kroger's headquarters.) The subsidiary conducted sophisticated explorations of checkout data and other information the retailer had linked to shoppers' loyalty cards, which were involved in nearly every transaction. The company likely also obtained personal information from outside firms. Kroger's 2015 privacy policy stated only that it would not sell or trade its customers' information; it did not mention gathering data from elsewhere.[59] Personalization was of course the goal. "Working with Dunnhumby," noted *Forbes* in 2013, "Kroger tracks each customer as an individual." A dunnhumby executive said the company strives to isolate each individual's behavior. "Do they have kids, do they skew toward healthy or fun, do they like organic or convenience, and where are they price sensitive—across all products or only on some." The firm uses this information and other research to understand its customers' movements through the store, and to tailor postal and internet mailings to individual shoppers.[60]

A number of grocery businesses, including Stop & Shop[61] and ShopRite, instituted a self-scanning-and-bagging program that enabled customers with the stores' app to monitor their purchase total as they shopped and then proceed quickly through checkout.[62] Supermarket and drugstore coupon dispenser Catalina Marketing built customized apps for this type of program to help its clients gather data on individual customers and offer personalized discounts. Now, instead of providing the coupons only through its checkout printers, Catalina could send customers messages as they scanned products while moving through the aisles. A brochure boasted that "Catalina Mobile targets more than 4.6 million personalized offers each month. Every offer is based on each customer's individual historical purchases, brand preferences and loyalty, location in store/aisle, and what's in their shopping cart. Catalina Mobile gives you the ability to target customers at the most critical step in the path to purchase."[63]

The program represented a major change. For decades barcode scanning and its link to rewards cards had taken place at the cash register at checkout, and now companies were hard at work to come up with ever-easier ways for customers to scan goods as they strolled past them. The Sapient Nitro agency developed a prototype involving a computer chip in the customer's shopping bag that would automatically identify the barcode of an item placed into it. The chip would then communicate with an app the shopper had downloaded to keep a running total of the transaction. Critics worried that customers might be less inclined to return to the store if they had virtually no direct interaction with a store representative—that is, the cash register clerk. They also worried that customers bagging their items in the aisle would encourage

theft (the industry calls this "shrinkage"). Although Catalina flatly stated that this was a "misconception," the company (along with others) nevertheless investigated approaches that would "mitigate the risks of bad behavior of a few customers, without alienating the majority of trustworthy, loyal customers."[64] One such tactic was a computer review of shopper scanning activity to look for an unusual number of voided transactions—an indication someone might have scanned an item and then pretended to return it to the shelf. Other potential red flags could be an extended period between scans (the implication being that an individual could be adding items to his or her shopping cart without scanning them), spot-checking whether someone has a history of being caught leaving a store with "accidentally" unscanned items, and what Catalina called "shopper status"—likely a reference to whether a customer's purchase history and loyalty status coincide with the items in his or her shopping bag.[65]

Beacon advocates asserted that their new technology would bring selling in the aisles to an even higher level of success. They said that BLE was superior to self-scanning-and-bagging programs because the shopper didn't have to handle a product to get a message about it or its competitor. Beacons were also quite inexpensive (by 2015 the price was down to $5 per box) and easily installed. When chain retailers such as Walgreens, Walmart, and Target saw how beacons could help them collect new behavior-centered data and send messages to identified shoppers based on extensive profiling, they pounced on the technology. The Business Insider Intelligence research group listed beacons as one of the most important new mobile developments helping merchants win back revenue "thanks to their ability to target consumers with sales and loyalty offers."[66] Critics, however, noted that beacon boxes covered rather small store areas and could not reliably follow an individual's complete movements through the store unless a large (and expensive) number of beacons were used. They also pointed out that smartphone users don't necessarily keep their Bluetooth phone signal turned on. A Kroger executive suggested that the firm didn't consider them cost-efficient; indeed, the company wasn't using beacons as of late 2015 and said it had no plans to do so.[67]

This chink in the Bluetooth box opened the door for alternative tracking solutions. IndoorAtlas, a Finnish company, uses a smartphone's magnetometer along with other built-in sensors to locate its position based on the Earth's "natural variations of geomagnetic field."[68] A retailer can then follow a phone that has its app fluidly through the store—with no need to invest in any boxes or lights.[69] General Electric is producing a system that incorporates beacon technology into a store's lighting system, so its range encompasses virtually the entire store. The GE system emits a unique light pattern to a phone if the user clicks OK to location awareness when asked by the store's app. The phone detects the Bluetooth signal specific to that location, and the phone's camera detects the light patterns from the fixture even if the Bluetooth in the phone is off. Noting one or the other signal (or both) allows the store to learn of the shopper's position and direction with an accuracy of less than three feet. The location information allows the retailer to send the phone a personalized message based on that location. And because interaction with the app allows the retailer to learn the phone's ID, the personalized message can be based on lots of other information about the individual in addition to location.[70]

Walmart was reported to have made a major purchase of GE's system.[71] Whether it or IndoorAtlas' approach gains wide acceptance, they reflect merchants' insistence on finding efficient ways to identify and follow shoppers in the aisles, sending them personalized messages in the process. Matthew Kulig, co-founder of the store mapping company Aisle411, believes that retailers ultimately will exploit a combination of solutions toward this goal. His company has been working with retailers to address both tracking and simply helping shoppers find their way around a store. Founded in 2008, the company began with a toll-free number providing computer-generated answers to questions about the location of items in stores. It has since evolved to using other technologies, including cellular, Wi-Fi, and Bluetooth, depending on the client. Perhaps the most advanced version of its products to date is a three-dimensional map created in cooperation with Google. Tested in some Walgreens locations in 2015, the system is accessed using Google's Tango augmented-reality tablet and provides product location information as well as customer tracking for presenting messages and coupons relevant to the shopper's profile.

Given the exhaustive efforts to target people's mobile devices inside brick-and-mortar establishments, it is not surprising that this technology began to extend outside the physical store. Facilitating this development was a GPS (global positioning system) chip that, by the 2010s, manufacturers were installing in every smartphone. The chip picks up the beaconlike signal of three geostationary satellites; software in the phone triangulates the data into map coordinates. The highest-quality commercial GPS receivers can pinpoint someone's position to better than

11.5 feet, and when combined with the location of Wi-Fi pings from stores and other places, the location can be even more precise.[72] Consequently, if a smartphone owner allowed an email provider, an app, or a website to access the phone's location, that information could be used to sell ads to merchants near that phone. The firms monitoring a phone's location could also—without additional permission—peddle the data to other apps or mobile websites, and *they* could sell the ads.

Predictably, advertising giants such as Google, Facebook, and Yahoo! became deeply immersed in GPS tracking, but so did inMarket and other smaller companies. inMarket, which provides location-based ads outdoors in addition to inside stores, offers merchants and brand manufacturers the ability to know where individuals carrying apps with its code are located on a proprietary latitude and longitude ("lat-long") grid containing a large number of retail locations. The ad-serving process works this way: If a person allows location detection on a smartphone, the phone's operating system continually pings the phone to check its latitude and longitude. The device then sends these coordinates to apps for which the device owner has allowed location. The phone essentially tells the app, "This is where I am. What should I now do with this information?" In the case of inMarket's apps or apps on which it has code, the answer is to plot the phone (and so the individual linked to it) on the company's grid. inMarket therefore knows where these people go from moment to moment. The company can also determine whether the phone holder is in an area where an advertiser wants to reach shoppers with a message. That space is called a "geofence"—that is, a defined geographic area. A retail establishment might pay inMarket to surround it with a geofence of a certain size so that when inMarket sees a mobile device enter, it will trigger a message about the store to be displayed through one of the apps with inMarket code. A geofenced area can be as large as a several-mile radius or as small as a single shop within a strip mall. In addition to targeting by location, inMarket can choose from among people in those geographically specific areas based on the demographic and behavioral information it gathered when people signed up for the apps, when it tracked their phone's movements in the outside world, and when it followed them in stores with inMarket beacons.

Another "geolocation" company, xAd, exploits this location technology somewhat differently. The company attempts to combine a smartphone's position with its user's actual reason for being at that location. xAd has access to both sorts of data because it provides a digital-advertising network for a raft of restaurant, tourism, and directory apps.[73] The advantage for xAd of working with such apps is that they logically require their users to accept giving out their GPS location to learn about establishments and their directions. Through these relationships, xAd claims the ability to know what the people in particular locations were intending to do. Based on close mapping of retailers' location coordinates, xAd could then serve ads within the apps that targeted "consumers as they are visiting key locations in real-time and showing their intent to browse and purchase in the physical world."[74]

xAd also uses the IDs of the smartphones it accesses to record physical locations along with analysis of buying intent. Simply knowing the past twenty retailers where a particular anonymous person shopped can provide useful information for target advertising. BIA/Kelsey senior analyst Michael Boland noted that "location always is thought of as geographic," yet it is most valuable "as a means toward building user profiles."[75] He said that many of the advertisers using these sorts of location targeting in 2014 were "the suppliers of consumer packaged goods products, including the Procter & Gambles of the world but also their retail affiliates like Walgreens." He noted that "they're kind of working together in ways that engage users to find the closest store."[76] At the same time, Boland said, xAd and other mobile targeters were applying their real-time capabilities to persuade shoppers to choose a client store over a competitor. Under such "geo-conquesting," as it is known, shoppers who are tracked near a competitor store are sent ads to persuade them to buy from the company's client instead. "For instance," wrote one mobile-marketing blogger, "a local coffee shop . . . might target users who are at big coffee chains like Starbucks and Dunkin' Donuts with timely messages emphasizing local deals, short lines or unique menu items . . . like 'Tired of waiting in line? Try our organic original roast today with no waiting!' "[77]

Thinknear is a company that also exploits location to target individuals, but in contrast with xAd it uses what's known as mobile advertising exchanges to reach individuals rather than specific sites or apps. A mobile advertising exchange is an electronic marketplace in which companies bid for the right to send advertisements to people from any of hundreds or thousands of sites or apps in near real time. Thinknear bids to serve ads for retailers and other clients based on what it learns through the exchanges about the mobile devices' locations as well as the demographics of their owner. The company uses GPS as well as Wi-Fi, cell triangulation, internet protocol (IP) address, and user registration to determine the location of the devices.

The retailing and marketing industries are also examining other avenues for exchanging information between a location targeting company and merchants. One possibility is to send individualized messages about a store to entice people when they are in proximity. If the person then enters the store, the targeting company would inform the merchant and share behavioral data. This activity could occur through such companies as LinkNYC, which began installing unrestricted Wi-Fi kiosks in public spaces throughout New York City in 2016. LinkNYC could collect data on the patterns of individuals who use the service as they pass the kiosks. The firm could then sell advertisers the right to send messages to the pedestrians based on what the kiosks had learned about their locations and behaviors. Scott Varlard, a technology expert at IPG Media Lab think tank, noted that the tracking of individuals could continue into the store if LinkNYC and the retailer collaborated on ways to match people's identities.[78]

As the capabilities to track and target individuals continue to extend their reach, it seems only a matter of time before the physical retailer can enter private homes. Of course, some versions of this type of interactivity already exist as people use their PC and laptop computers, tablets, and phones at home to make purchases or to gather coupons and compare prices in anticipation of physical shopping. Not too long from now, insisted David Skaff of the Science Project marketing agency, there will exist what he calls "omnichannel nirvana": interactions of common appliances, including thermostats, lights, and refrigerators, with each other.[79] The connected home, part of a larger development technologists call "the internet of things," has been leading retailers and brand manufacturers to revel in the possibilities of getting in on the stream of habits, lifestyles, and even personalities that the collection of household-use data would reveal. Leigh Christie, a technologist at the Isobar agency, predicts that people will "shop on" their family members and friends, meaning they'll be able to use a device to scan their clothes to obtain information such as where the items were purchased. He says that food containers will be able to remind their owners verbally of their expiration dates, and where they can replenish them. He also believes that everyday products such as cosmetics will reach out to individuals with advice relevant to the season and the weather.

Such communications are feasible today for cosmetics, Christie notes (though with Android phones only; iPhones don't as yet allow this sort of interconnection). A cosmetics manufacturer or retailer can place an inexpensive chip (known as a near-field-communication, or NFC, chip) on an item—for example, on a cosmetics container. The chip will send its unique ID code to a smartphone's NFC transmitter/receiver when the two are in very close proximity. Phones on which the cosmetic firm's or the retailer's app has been installed will receive the ID code, which identifies the product. Alerted to the specific cosmetic, the app can then instruct the phone to carry out a variety of tasks, such as looking up the weather and the time of year, and checking previously collected information about the phone's owner in the cosmetic company's computers. The app can then send a personalized message to the phone based on the information it has compiled—for example, "It's a good day to wear blue," "Your red lipstick would go great with this makeup this time of year," or "Here's a coupon for a new look!"[80]

* * *

Strong tensions shoot through this evolving world, as competition among companies continues to intensify and executives, consultants, investors, vendors, and advertisers try to assess what the role of mobile-tracking technology should be. A number of executives suggest that many of the so-called crucial developments in tracking technology are in fact anything but. Some even consider BLE beacons, the shopper-hunting stars of the mid-decade, to be merely an intermediate stage that will ultimately be cast aside for something different and, hopefully, better. At the same time nearly all agreed that the basic imperatives of identifying and following shoppers to and through the physical store, collecting enormous amounts of information about them without their awareness, and personalizing messages to them along the way is, whether through one technology or another, a new requirement of retailing.

More pressing than the technology itself is the key coin of the realm: data. Who owns customer data when so many parties, often in collaboration, ping and poke at shoppers at home and outdoors, and in stores? inMarket executives voiced specific concern over the lack of control their retailing clients feel they have over who should be allowed to reach shoppers within their walls, and by what means. inMarket contends some brand marketers have begun putting their own detection devices into store displays as a way to audit the number of passing shoppers and identify their app holders, and even to serve coupons to these shoppers. Marketers assert that these activities are always undertaken with the permission of the retailers, but inMarket officials dispute this and note that they have undertaken measures to thwart unauthorized activity.[81]

Ivan Frank, the digital-marketing head of the upscale Taubman mall company, agreed that people shopping in one of the company's malls should not be receiving advertising from outside firms, potentially encouraging them to buy items online instead of at the mall. Although shoppers can't be prohibited from accessing Amazon-like apps or getting text messages from retailers as they browse, he said he was working to at least maintain control of beacon companies hired to install units in the malls—but this was proving difficult. For example, some of the beacon companies that Taubman was planning to hire to set up and run units in a new mall the firm was opening in Sarasota, Florida, wanted to own the data their beacons collected on customers moving through the mall. "Why," Frank asked rhetorically, "would we want a third party to own our data?" He objected that this information could be sold to retailers competing with Taubman's tenants and mentioned Amazon in particular—"a company we're at war with." He lamented the lack of clear industry norms concerning beacon companies' practices.

Frank's concerns are shared by others in the industry. inMarket's carefully spelled-out hierarchy of data-sharing among its retail and app partners is being accepted by many. shopkick and Catalina typically send the names, addresses, and email addresses of all their shoppers to each of their "partners," but they do not release other individual shopper information—for example, the items bought or returned, the amount spent—except to the retailer that made the sale. Of course, firms such as shopkick and Catalina themselves have access to a range of data about individual shoppers from many different merchants, and the merchants try to negotiate limits on the amount of information those firms can store and on how they can use it.

As retailers, technology companies, and brand manufacturers negotiate new ways to glean information about people and implement marketing efforts based on that data, the one party who has no say in the matter is the shopper. Although privacy policies do allude to the often voluminous amounts of data retailers can ingest about people, studies repeatedly show that relatively few consumers read them. Those who do likely find the policies' legal and industry language nearly impenetrable. And if they do manage to wade through the policies and object, the only option is not to use the website or the app—the rules typically constitute a "tough luck" contract, not open to negotiation. Shoppers who are determined not to give up their data can find ways to stop Web trackers and block ads, and although an increasing number of Americans are doing so, most remain readily identifiable and trackable via various cookie-like tags and data-matching procedures on the Web and on mobile apps. Physical retailers, though, have particular reason to worry whether shoppers will accept in-store tracking because, unlike digital retailing, crucial aspects of it cannot as yet be made as invisible. Their approach instead is to get shoppers to accept, and then take for granted, the omnichannel world of tracking and targeting, from their home to the brick-and-mortar store. To do so the merchants must seduce shoppers to switch on their Wi-Fi and/or Bluetooth, download the retailers' app, and set up their loyalty program, thereby inviting the discounts and other blandishments offered up by the newly outfitted physical store. This evolving strategy is ushering in a new battlefield for shopper loyalty.

5 LOYALTY AS BAIT

Airlines are an extreme example when it comes to discriminating among customers for profit. Over the years, frequent-flier reward programs, which were launched to entice business travelers to book all their flights with the same airline, typically have become increasingly complex. Carriers created multiple tiers within their programs, offering their most frequent fliers the best perks such as preferred seating, more legroom, early boarding, and guaranteed overhead luggage space. Lower-tiered fliers and passengers who haven't joined the frequent-flier program aren't treated nearly as well; for example, they typically receive less-comfortable seat assignments unless they choose to pay extra. The airlines give business travelers, who are their main targets, enough protection and privilege even at middle loyalty tiers to make them oblivious to the indignities experienced by infrequent fliers. Indeed, the *New York Times* noted that "the opposite of rewarding a good customer is penalizing a bad one." In the case of Delta Air Lines, the paper wrote, "in order to toss 10 or 11 [reward] miles at some travelers [its best customers] for every dollar they spend, Delta will have to do less for plenty of others on many of its routes."[1]

The airline loyalty programs have had enormous influence on marketers' overall efforts to use data to attract, learn about, and reward the best customers. Consultants have exhorted merchants outside the airline industry to discriminate between customers whose purchases indicate they will bring the firm profits over a number of years (a high lifetime value) and the others who just show up. Bain & Company loyalty consultant Fred Reichheld wrote in a letter to the editor in *Harvard Business Review* that "plenty of long-term customers—those who relentlessly cherry pick loss leader products, who treat front-line employees abusively . . . can't be considered loyal." He added, "If frontline workers are taught to distinguish profitable from unprofitable customers, and if those employees are rewarded for focusing jointly on earning profits today and treating profitable customers in ways that make them want to return, companies will be pleasantly surprised by the resulting improvements in growth and profits."[2]

Several retail chain executives interviewed for this book suggested that the bluntness with which this sorting activity takes place in stores depends on the merchant's target audience. Although luxury brands such as Louis Vuitton, Bergdorf Goodman, and Saks might well want the public to see them as exclusionary in order to cultivate elites, mid-tier merchants cannot be perceived as treating their lower-value customers the way that the airlines do. One department-store executive said he would rather try to persuade customers to move up the purchasing chain instead of making them uncomfortable or implying they should shop elsewhere. An executive with a major regional supermarket chain shares that view; rather than make life difficult for less-profitable shoppers, she said, her company tries to convert them to becoming increasingly loyal and, consequently, more consequential customers for them. "Cherry pickers exist," she said, "but we don't fire them. They are customers. Groups of folks in the company deal with such things, worry about loyalty now and beyond. There are certainly different types of programs [to reward different levels of loyalty]. If you use your [basic supermarket loyalty card] you will get the base level of savings. But if you use other programs, like gas . . . that can bring the customer more benefits."[3]

The key to selling, executives agree, will increasingly be to know how to deal with individuals. Merchants need to know what messages to send to specific customers, whether store associates should interact with them (and, if so, for how long), and whether the store should offer specific customers discounts and, if so, for which products and for how much. To do this the store must have appropriate information about each shopper each time that person enters a digital or physical version of the store. "Retailers know they need to upgrade their technology portfolio to understand consumers' complex path to purchase," said a retailing consultant in 2014 at the National Retail Federation conference. "It's critical for them to reach customers in new ways, and that can only happen if retailers gather and analyze customer data." Macy's chief marketing officer put the matter more bluntly. "We don't need more customers," he noted in 2012. "We need the customers to spend more time with us."[4]

He meant that the merchant was attempting to engage shoppers in every way possible—the internet, any mobile device, and in the physical store, and he was particularly referring to the value of shoppers who interacted with more than one of these channels, as they spend more than those who adhere to just one. In an article discussing Target, the *FierceMobileIT*[5] trade journal noted that "the multi-channel shopper is the prize, as customers who shop both in stores and online generate three times the sales compared to shoppers who only frequent brick-and-mortar stores, according to the retailer." Target, Macy's, and other chains have recognized that identifying and pursuing such high-value targets require highly specific data, and lots of it. "Analyzing data can help companies make precise decisions about everything, from crafting relevant offers for on-the-go consumers to displaying the right assortment of merchandise," noted a retailing journalist who was covering the 2014 National Retail Federation conference, which was packed with merchants with strong physical footprints and digital sales platforms.[6]

The increasing focus on the individual was by no means unique to retailing. "Companies across the globe and from every industry are building teams to translate mountains of information into better understanding of their customers," stated a 2014 Forrester report titled *The Age of the Customer Requires a More Intelligent Enterprise*. The title alluded to the potential knowledge and bargaining power shoppers could exercise in the age of the Web and mobile phones. "Your customers' expectations are changing," the report cautioned, "fueled by the constantly connected world and the mind-boggling speed of technology enablement," and companies have to respond in kind. "Knowing your customers and quickly translating and applying that knowledge across the entire enterprise is no longer a competitive advantage, it's a competitive necessity if you are to win, engage, and retain customers." An "intelligent" business, it said, is one "in which customer knowledge is drawn from everywhere, created centrally, and shared across the entire enterprise, so all stakeholders can act upon it and measure the results." Key to this approach is creating a unique identifier for each customer to "enable firms from industries as diverse as telecom, financial services, media, retail, and high-tech to avoid the agony and inefficiency of manually tracking, trending, and tying together customer interactions from disparate databases." All this information should tie into "an automated system built on rules-based algorithms that trigger actions such as personal communication, offer optimization, and cross departmental notifications." Such automation would speed the time it took to translate executive insight into action so as to "meet customer expectations."[7]

A central question remained: What are the best ways to get such information without alienating customers? Although marketers had come up with surreptitious ways to create shopper profiles, not all of these would work in a selling situation. Consequently, many physical retailers began to use their loyalty programs to accomplish it. Traditionally stores used the programs to generate recurring purchases; now they also began to use them as bait to attract shoppers and encourage them to consider it perfectly normal and ordinary to give up information about themselves unselfconsciously. The shift has ramifications far outside physical aisles.

Retailers in the early twenty-first century are adopting a new perspective in customer analysis that goes beyond traditional classifications of people based on a small set of categories such as demographic (gender, annual salary, etc.) or lifestyle (hunters, skiers, people who eat out a lot, etc.). Though these classifications remain important, there is a new emphasis on gathering behavioral and attitudinal information in order to predict how specific people will respond to various shopping situations. Indeed, practitioners in fields such as advertising, health care, political analysis, and insurance have been increasingly accepting that learning as much as possible about individuals will lead to accurate predictions of their behaviors. Genetic research has contributed to this new mindset. When the Human Genome Project published the first human DNA blueprint in 2000, "scientists promised that the age of personalized medicine had arrived."[8] Although the technology has yet to deliver on this promise, hope remains that improved whole-genome sequencing techniques will ultimately usher in a new health care era. A *Los Angeles Times* article noted that "what they learn will enable doctors to warn their patients of their genetic vulnerabilities, allowing patients, in turn, to take steps to reduce their risk."[9] At the same time, firms such as AncestryDNA (the genetics division of Ancestry.com) reflected and encouraged public fascination with genetics in their marketing efforts. "This service combines advanced DNA science with the world's largest online family history resource to predict your genetic ethnicity and help you find new family connections," explained the AncestryDNA website.[10]

Retailers adopted geneticists' fixation on the individual but focused on more realistic avenues for collecting data. They turned to the work of actuaries, statisticians, accountants, engineers, and computer scientists who were building disciplines aimed at calculating the probability of knowing individuals' actions based on social characteristics (such as age, ethnicity, gender, race, membership in an organization) and behaviors (such as driving habits, web-surfing habits, vacation destinations). A century ago the concept of probability "was relevant to practically only two types of people: to professional gamblers—and actuaries. Another profession which has integrated the concept of probability increasingly since the middle of this century is that of accountants." But in recent years, "the concept of probability has become a universal way of thinking which is incorporated in all aspects of our life."[11] By 2010 computer-analytics specialists had attached that idea to tracking the minutiae of people's everyday activities—"the little data breadcrumbs that you leave behind you as you move around in the world," as MIT computer science professor Alex Pentland put it.[12] Catalina Marketing used a different metaphor: "Like fingerprints, every shopper's profile is unique in the assortment of products they buy each year."[13] The work of zeroing in on patterns linked to the fingerprint bread crumbs typically involved complex mathematical tools with labels such as data mining, logistic regression, and cloud computing.[14] It attracted many different types of companies. A 2011 study forecast their phenomenal growth, including the need for 1.5 million managers and data experts by 2018.[15]

A 2013 report from the Gartner information technology consultants celebrated the rise of this analytical class. It predicted that within five years businesses would be able to identify and follow individuals through apps on mobile phones, home appliances, cars, and wearable devices. Gartner identified four stages in the evolution of these activities: "Sync Me," "See Me," "Know Me," and "Be Me." Sync Me and See Me "are well underway," it said, referring to the increasing tendency of customers to connect several devices and the increasing ability of businesses to follow them across these technologies. Surveilling people across their digital and physical worlds would lead to such in-depth profiling that businesses would create intimate models of an individual's behaviors and dispositions regarding their consumption patterns and preferences, and would ultimately enable firms to anticipate customers' needs and actions.[16] The report added that when a shopper can be identified via the information gleaned through connected devices such as location, app-usage tracking, customer demographic data, and behavior history analysis, "it will be easier to target the right advertisement at the right moment, or send the precise redeemable coupon when people are at the right location, at the right time of day, generating a higher return on marketing campaigns than before."[17] The report seemed to echo Google chairman Eric Schmidt's comment that the information Google has collected on individuals enables the company to know "roughly who you are, roughly what you care about, roughly who your friends are"—to the extent that it can give insightful suggestions about what individuals should do next in specific locations.[18]

As to the challenge of collecting customer data, the internet set a precedent that physical retailing is now trying to imitate regarding the tracing of customer behavior. From the dawn of Web commerce in the mid-1990s marketers learned a lot about shoppers, and mostly without their direct approval or knowledge. The basic Web tracking mechanism, the cookie, was developed in 1994 solely to address the issue of identifying individual shoppers so they could purchase more than one item in the same transaction. Marketers learned quickly that such a mechanism could do much more, and an entire industry has since grown up around the ability to tag people, follow their movements from website to website, sometimes connect them to personally identifiable information (name, address, phone number), and build information profiles about them based on their digital travels and, often, their offline lifestyles.[19] People have willingly provided personal information about themselves by creating accounts at sites of interest, but most of the time they have scarcely considered the ramifications of sharing such data. Surveys show that Americans have had little understanding of the specific ways in which tracking takes place or of the general lack of government regulations overseeing such surveillance.[20]

Americans' ignorance has been no different when it comes to the mobile phone. Especially after the iPhone's 2007 release, advertising and retailing executives considered it a necessity to track people via their smartphones in addition to following them on their home computers—and to pursue them from one device to another. However, a cookie cannot follow someone across platforms; it can't even follow someone using different browsers on the same platform. In addition, while a smartphone's Web browser such as Safari or Chrome can accept cookies, apps cannot. Marketers were flummoxed. "When companies lack a central notion of customer identity, each channel and the system creates a new identity," noted a report from the Janrain marketing consultancy. "When identities are [isolated], companies all together lack the ability to plan, optimize, and measure the customer journey across devices, channels and systems."[21]

The best solution seemed to be what data scientists call the deterministic approach, meaning the shopper has to register separately for both the website and the app. Facebook and Target are businesses that operate this way. Say a Target customer logs into the company's app and website every time he or she uses a device (or the software remembers the person). The tracking, the analytics, and the profiling can then take place in the background, out of the person's awareness. Under this system, the company would recognize the individual no matter which of the person's devices was used. Even if a shopper doesn't sign in on every device, it might be possible to determine the person's connection to all of them by encouraging the individual to offer personal information when visiting a website or using an app on a device not yet linked to that individual. As that personal disclosure takes place, the merchant can make a permanent connection between the shopper's identity and the device ID. A common tactic for unmasking people, according to Janrain, is "to offer a discount code on a next purchase if an anonymous site visitor would provide an email address and opts in to future email marketing campaigns. Consumer brands frequently run sweepstakes and other contest promotions to the same end, and many product manufacturers incentivize otherwise anonymous buyers to provide contact and demographic information through warranty registrations."[22]

But what if a retailer wants to learn about people who have never logged in to its app or website? Beginning in 2007, analytics firms launched a cavalcade of new technologies to follow these and all other individuals across devices, often without their being aware of it. The approaches they use are called probabilistic (as opposed to deterministic) identification because when firms don't have log-ins to work with, they have to use predictive methods to recognize when a person using one device is the same one who is using another. Tapad is a company that does that. It analyzes the IDs and IP addresses of millions of devices to identify their geographical locations and then looks for patterns: are a laptop, tablet, and a smartphone using the same internet connection? Are the web-browsing patterns similar? Do the devices "wake up" at the same time? Do they "sleep" at the same time? The greater the number of "yes" answers to these questions, the more likely that the devices connect to the same user. Although accurate analysis might seem improbable,[23] independent tests suggest a higher than 90 percent accuracy rate in predicting common owners.[24]

Tapad also matches its tracked phones, tablets, laptops, and desktops with devices that database firms such as BlueKai and eXelate follow with cookies and other tags. Tapad purchases information from these companies and uses all the information it has gathered to create individual profiles, which it then sells to marketers for targeting with personalized messages.[25] Surprisingly, Tapad doesn't try to match device owners with information that can identify them offline, such as name, home address, phone number. It also doesn't permit clients to use its technology for tracking people's activities within areas of less than one mile or to send people ads based on precise locations without their explicit consent. "Time to get precise (without getting personal)," the firm tells potential clients on its home page.[26]

In contrast, plenty of cross-device tracking firms do help retailers and other marketers connect behavioral data with personal identities. Janrain does this through social network log-ins, in which people register and sign in to a website or app with their passwords from one of the social networks to which they belong. Janrain claims that "more than half of people" worldwide use their social network log-ins for this purpose, with 81 percent going through Facebook or Google+. This method is certainly convenient, as people don't have to take the time to register on sites or apps requiring users to set up an account; the social network password enables them to bypass the process altogether. But in so doing, the social network logging them in can record this activity. Moreover, these "social log-in identity providers," like Facebook and Google, enable the site or app owner to identify each person by name as well as "select specific pieces of customer data that go beyond the basic name and verified email address to include elements like a customer's birthday, Likes, relationship status, photos, and friends' lists, to name a few." And the social network updates the data every time a person logs in again. "Your customers simply select their preferred social identity from Facebook, Twitter, or other networks," noted Janrain in its promotional literature, "and choose whether to share some of their information with you." Logging in to the same retailer on various devices using the same social log-in presents a persistent identity that enables retailers to compile extensive amounts of information on individual shopper behavior.[27]

In recent years privacy advocates have convinced Facebook, Google+, and other social platforms to notify their users about the personal data they give away when they log in with their social accounts. In 2015 Facebook began to allow members to specify data they don't want to share when logging in to a site or app. A Janrain report noted that "some consumers are discouraged from using social log-in when the personal data requested seems excessive for or irrelevant to the intended transaction." To allay such concerns, it said, the retailer should ask the social media site to transfer only the data that will allow the retailer to recognize the individual. But once that basic material pertaining to identity has crossed the threshold, the data floodgates would be open. At that point, Janrain advised retailers, "build a supplemental strategy to collect everything else [about each person]."[28] The company listed a range of sources that merchants can consult to collect "everything else," including email service providers; gaming, e-commerce, and online chat and commenting platforms; "social signals/social listening" services such as Twitter, Facebook, and YouTube; rating and review platforms for those who offer opinions on products, services, or companies; and on-site behavioral analytics.[29]

Retailers who want still more information about their customers, including statistics that can potentially identify shoppers on their sites and apps on different devices, can buy it from data brokers such as Acxiom and Experian, which by the mid-2010s had put together their own ways of tracking individuals across digital devices. In 2013 Acxiom introduced a system that gathers information continually on approximately seven hundred million identifiable individuals using three sources: Fortune 100 companies' records of people who "purchased something or signed up for a mailing list or some kind of offer"; "every data attribute you could gather publicly about consumers from public records"; and its own cross-device cookie-like system that can identify an individual across many digital properties and gives Acxiom "access to the last 30 days of behavior on more than one billion consumers."[30] Acxiom executive Phil Mui claimed that "for every consumer we have more than 5,000 attributes of customer data." Joe Mandese, editor in chief of the online industry publication *MediaPost,* asked Mui and Acxiom CEO and president Scott Howe about the seemingly "creepy" nature of the company's aggressive quantification of individuals. He reported that the two executives viewed their work as "just a fact of modern consumer life, and all Acxiom is trying to do is make it more scientific so that the data works the way it should, friction is taken out of the process for marketers and agencies, and consumers—at least—get the most relevant offers and messages targeted at them." Indeed, Mui boasted that Axciom can actually predict individuals' future shopping behaviors based on the demographic information and the off- and online purchase data the company has compiled. "We know what your propensity is to buy a handbag," he said. "We know what your propensity is to go on vacation or use a loyalty card."[31]

Despite the enormous and still-growing capability of companies to gather data from various devices so they can identify and market to consumers, retailers remained concerned about accomplishing the same with shoppers strolling through the aisles of their physical stores. These shoppers have to be explicitly willing to be identified if a human sales associate or a computer-driven point-of-sale machine is to recognize them and follow their actions. There is also the matter of encouraging customers to make sure the location tracking (Bluetooth, Wi-Fi) functions are activated on their phone, as well as to download the retailer's app. High-profile privacy issues concerning National Security Agency surveillance and "creepy stalking" by marketers have prompted many Americans—approximately 35 percent, accordingly to a 2013 Pew Research report—to turn off their phone's Bluetooth capability.[32]

Retailers argue people don't have the antipathy toward tracking that critics claim they do. Some observers have declared the situation a "privacy paradox"[33]—meaning that, the *New York Times* noted, "normally sane people have inconsistent and contradictory impulses and opinions when it comes to their safeguarding their own private information."[34] A March 2015 Accenture consulting firm survey found that people felt strongly about controlling access to their personal information. Fully 90 percent of the survey participants said that, if given the option, they would limit access to certain types of personal data and would stop retailers from selling their information to third parties. In addition, 88 percent wanted to be able to approve how their data was used. But Accenture also found contradictions among the people in its sample: while nearly 60 percent of consumers wanted to receive promotions and discounts, just 20 percent approved of retailers knowing their current location so that the retailers could tailor those offers.[35]

An Accenture representative underscored the importance of gathering data about shoppers despite their reluctance. "Addressing the complex requirements of U.S. consumers is challenging because they are conflicted on the issue," he noted. Yet he went on to state that "personalization is a critical capability for retailers to master." The solution, Accenture suggested, lies in persuading shoppers to see the situation in terms of trade-offs. "If retailers approach and market personalization as a value exchange, and are transparent in how the data will be used, consumers will likely be more willing to engage and trade their personal data."[36] In reality, large retailers have yet to express such an interest in transparency and assume instead that their customers want personalization and will willingly exchange their data for relevant offers and good buys. Forrester marketing consultant company mirrored this view in a report that suggested ways to reach goals regarding shopper data but didn't mention a need to let people in on how the information about them would be used. Forrester offered its own pragmatic formula: "Provide services that customers find useful and get back data on product use and customer affinities."[37]

Physical retailers see cultivating loyalty as a way to encourage customers' allegiance to services they consider useful. Retailers crave the kind of allegiance that will compel customers to identify themselves and to turn on their Bluetooth or Wi-Fi when they walk into a store because they think these actions can help them even if they are not sure exactly how. More generally retailers hope loyalty will trump wariness so that shoppers will suppress any data concerns and happily give out information about themselves—and will do so across multiple platforms, and will offer far more facts about themselves than they might even realize. Retailers especially hope they will be able to reinforce this relationship with the most desirable of those customers, encouraging them to implicitly accept and trust what the company does with all their personal data.

The intense competition accompanying the rise of online selling helped catalyze retailers to create "loyalty" programs that offer incentives for customers to return. Whether the programs retailers set in motion to accomplish this goal truly constitute *loyalty* depends on one's definition of the term. Many have questioned whether this word accurately describes a customer's repeated purchases at a store to accumulate points toward a gift. *Bribery* might be a better term, though the amount returned to the customer is typically quite small—often just 2 percent. In academic circles some have explained a person's repeat visits to a store in response to loyalty points as a merchant's attempt to imitate the behavioral psychologist B. F. Skinner, who conducted experiments with animals using food as a reward to encourage specific behaviors. To apply this methodology to retail shopping, the customer learns the habit of returning "first through short-term 'points pressure' and then through long-term 'rewarded behavior' that results from the reinforcement of behavioral learning by gratification" from those loyalty points.[38] Many agreed that a successful stimulus-response regime could encourage loyalty, but they did not think that the goal should merely be to get customers to return. After all, buying things might have little to do with adding value to a retailer's bottom line; for example, customers who buy only deeply discounted sale items provide little benefit to a retailer.

Bain consultants Fred Reichheld and Phil Schefter believe that loyalty must involve focusing on "repeat purchases among a core of profitable customers."[39] Retailers tend to accept this view, yet they typically consider emotional attachment to be a crucial aspect of customer loyalty. Emilie Kroner, who heads consumer markets organization engagement at dunnhumby, a retailing consultant, observed in 2013 that "emotional loyalty to a brand is becoming more and more important" to merchants.[40] Bond Brand Loyalty, a creator of loyalty programs, declared in 2014 that "it's time for brands to look beyond [customers accumulating rewards] points to establish deep, meaningful relationships, even bonds, with their customers in ways that are engaging, emotionally rich, and brand aligned."[41] Such affective links are important, Bond Brand suggested, because they can help facilitate the collection of individual customer data: emotional loyalty results in higher trust and, consequently, a willingness on the part of program members to offer up personal information. Marketers then send out personalized messages built on this data, resulting in "likely . . . incremental business results."[42] Some believe that engaging shoppers in games as part of loyalty programs (what the industry calls "gamification") is effective in encouraging emotional connections. Ayan Sen, senior manager for the marketing consultant company Peppers & Rogers Group, suggested that shoppers value this aspect of rewards programs nearly as much as they do protection and prestige. "Gamifying the engagement platform helps customers enjoy the activities that earn them rewards," Sen wrote. "With a typical tiered loyalty program, many consumers don't get close enough to the top rewards, so they lose interest."[43] The president of loyalty marketer Points agreed, noting that the aim of "modern gamification" is to create "a relationship of exchange where consumers are challenged and incentivized to share more of themselves: their time, their attention, their information, and ultimately their loyalty."[44]

Merchants with physical stores see emotionally driven loyalty programs as a way to address their need for in-store data collection in the face of widespread public opposition toward surveillance, offering program members protection, privilege, and games in exchange for willingly providing the store with information about themselves. Retailers and their consultants believe that the data collected in this way allows them to create increasingly relevant messages. Wanting relevant messages, shoppers will accept location tracking and keep on Wi-Fi and/or Bluetooth and/or offer up their mobile phone numbers. Under this same relevance-through-personalization banner, the stores can make their customers complicit in their own data releases without describing the particulars of profiling and differential treatment lest the customers get annoyed or balk. It's a classic technique of misdirection: hype the benefits of a relevant brand experience using the traditional emotional carrots of privilege and protection. Garnish it every now and then with the fun of gaming that in-store technologies can offer. And quietly, even surreptitiously, gather, store, and analyze all the data that go along with these activities.

The Janrain consultancy suggested to retailers in a report that they offer loyalty members the privilege of "first-to-know" promotions and other members-only offers not only to encourage shoppers to identify themselves, but also to motivate them "to return again and again and offer up more first-party data or engage in behaviors that help brands identify them as advocates or influencers."[45] Discrimination was an integral part of the process. "Not all customers are created equal—and not all segments are, either," the company noted in a second report. "Segmentation should always be in support of maximizing customer value and response rate, and ultimately return. If your analysis of customer identity data reveals segments that are disengaged, or have never demonstrated a target behavior or conversion, focus on creating compelling content for your other more high-value segments first."[46]

Cosmetic chain Ulta Beauty exemplifies the new approach to customer loyalty programs. Founded in 1990, the company currently operates nearly eight hundred stores nationally, selling both inexpensive and high-end brands. According to an analysis cited in *Investor's Business Daily,* Ulta's target customer in the mid-2010s is an upscale woman in her early thirties, with a household income of about $75,000. The company has grown steadily, by approximately 10 percent each year, and its success has been attributed to many factors.[47] An article in *Investor's Business Daily* suggested that customers are drawn to Ulta particularly because of its leisurely ambiance. "Salespeople aren't on commission so they're not breathing down your neck," the article said. Another analyst attributed Ulta's success to aggressive promoting of high-brand awareness, including its marketing efforts on Ellen DeGeneres's television show. "A loyalty program, new products and the perception that it's a 'destination shop' help pull customers to the store," he asserted. "They know how to incubate new brands through their merchandising program. That's what women are looking for. They're looking for 'the new thing I can show off to my friends.' "[48] Another pointed to Ulta's one-stop shopping experience. Before Ulta, he noted, women had to travel around to beauty stores, department stores, and mass retailers to make all their makeup and fragrance purchases. He suggested that Ulta allows them to short-circuit that journey, even to the point of having salons at the back of its stores that offer hair, skin, and brow services. He also contended that Ulta can hold its own against Amazon's online "Beauty Bar" collection because "most women go online to see what's hot but prefer going to brick-and-mortar stores. They want to see the makeup and smell the fragrance." A Wells Fargo prognosticator remained unconvinced, however, that a niche physical store can be adequately protected from the two nemeses of retailing. "At the pace at which Amazon and Wal-Mart are innovating, if you're great you've got to be supergreat tomorrow and unbelievably great a day later."[49]

One way the chain attempts to remain competitive in the new shopping universe is to offer thousands of its products on its website, where shoppers can also engage in live, interactive chats with fashion experts, see photos of celebrities using Ulta products, and view photos as well as so-called haul videos of products customers have purchased.[50] (Many of these features, especially the photos, are also available on Facebook, Pinterest, Twitter, and Instagram.) Although the company promotes these extras to add credibility, authority, and authenticity, its larger aim, noted Ulta's chief marketing officer, centers on enticing high-value customers. Like many retailers, the chain has found that this group encompasses shoppers who cross the chain's virtual and physical boundaries. "Our best guests," the marketing chief said, "are the ones that are using our physical stores as well as our e-commerce business and then ultimately using our [salon] services—a trifecta that works well for us."[51]

That "trifecta"—along with its associated marketing tools involving smartphones, guest services, direct mail, online and in-store activities, email, and social media—is the fundamental source for the data that fuels Ulta's "Ultamate Rewards" loyalty program. It encompasses a fifteen-million-member database that contains customers' unique preferences so the company can "target the conversation" to individual shoppers. Individuals can become members through signing up directly for an account or logging in through Facebook or Google+; these social log-ins are a particular data boon for the retailer because Ulta might then be able to learn the names of the member's friends. On its website Ulta emphasizes that membership in its reward program "has its privileges." The program's rules are a bit tricky, and they have changed more than once. Not surprisingly, they are set up to encourage customers to spend more when they make purchases. The description glosses over the rather small standard discount the program offers and instead encourages members to earn double and triple points on specific products "or on your entire purchase during special bonus days." If shoppers spend $400 within a calendar year, their points don't expire. Moreover, in mimicking airline loyalty programs, customers who reach this "platinum" status receive a higher reward ratio: 1.25 points (rather than 1 point) for every dollar spent. Consequently, this group can receive discounts more quickly than lower-spending members, and it aims to encourage high-value customers to spend to keep their special status.

The retailer's additional goal of learning as much as it can about customers and reaching them across digital and physical domains may or may not be evident in the postal mailings, emails, app coupons, website offerings, Facebook blandishments, and other marketing materials program members encounter. Members are probably completely unaware of other aspects of data gathering that are explained in the small print of the Ulta privacy policy. If they do take the time to wade through these notices they might be startled to learn that Ulta:

- claims the right to buy information about its members from third-party sources (often database firms) to incorporate into its existing member profiles;
- uses cookies and similar elements to follow members' activities on its sites and in emails as well as to infer their interests by tracking members' visits to non-Ulta sites;
- allows authorized advertising networks to collect data about its members while they're visiting the Ulta site;
- tracks members' locations via their mobile devices without requesting explicit permission (unless members turn off that feature on their devices);
- collects members' comments and other input regarding "a product review, question, or other information" on the Ulta website; and
- allows "third-party social media widgets such as buttons or similar mechanisms" from Facebook, Twitter, Pinterest, YouTube, and FourSquare to collect information about members as well as to place trackers on members' devices.

Although Ulta collects much information about its shoppers online and via its app, Ulta's marketing director emphasized in 2015 that his company's goal is for them to shop primarily in the store, and to have "a personal and fun experience" as they do it.[52] To help facilitate this, the company has instituted what it calls a "clientelling" tool—an iPad that provides clerks with immediate access to every bit of information the store has compiled on any individual customer. Once the employee has identified a customer, the clerk can "review her purchase history, understand what her beauty concerns are, and target the right product to address those concerns." The associate can also enter the customer's brand preferences. All that information can lead to personalized offers as well as "targeted invites" to in-store events.[53]

Sephora, an Ulta competitor, which likewise collects information on its customers from their activities via the internet, mobile devices, and in-store, has a different focus. It installed beacons in all its locations in 2015, and smartphones that have downloaded the company's app can capture the beacon transmissions, which send notifications of promotions in specific areas of the store. The app also outlines in-store services, and shoppers can also use it to review previous purchases as well as to check their current balance for the company's rewards program. In-store shoppers who have checked their balance to find that they're just a few dollars short of earning their next reward often shop accordingly and buy a bit more to make sure they're eligible for their next reward once they complete that day's transaction. Johnna Marcus, Sephora's director of mobile and digital store marketing, said that the balance-checking function of the app is one of its most popular features.[54]

No matter the merchant, rewards programs have much in common with game playing—following rules, accumulating points, obtaining prizes. To be sure, industry reports suggest that such game-like activities can help keep customers interested in the merchant—"prime the loyalty pump," says an executive at the marketing consultant Pepper and Rogers—by informing them about products, stimulating their use of more services, and encouraging them to talk about the company with others through social media.[55] For the retailers involved, the fun elements help sew together each customer's purchase and preference data from the merchants' digital online, mobile, and physical platforms. In 2015 Sephora extended the game-like approach beyond its rewards program to an in-store "augmented reality image scanning" activity. Customers who spot "Scan to Watch" signs in the store are urged to point their phones at the signs and "unlock content (such as exclusive videos)."[56] The signs also encourage shoppers to use the company's app as a tool as they shop—which also means that the store's customer-tracking beacons will be capturing data on its customers as they shop.

While many, if not most, loyalty programs can offer real value to shoppers, their unstated aim is also to train people to give up personal data willingly. shopkick, which pioneered ultrasonic and Bluetooth technologies to follow customers through stores and encourage them to interact with the retailer and provide personal information, has made capturing all sorts of data with little transparency an integral part of its business model. In an uncharacteristically frank reflection, a Best Buy representative said her company signed up with shopkick to "find new ways to bring people into our stores through the use of mobile technology, but more importantly, [to] know when they [individually identified shoppers] come into our store."[57] For shoppers shopkick emphasizes the rewards they can earn by scanning goods with the shopkick app and by purchasing products. According to the online magazine *TechCrunch*, an important part of shopkick's success came from introducing "a level of 'gamification' to the shopping experience." But the company's privacy policy makes clear that it reserves the right to store, analyze, and use for profiling and "tailored" targeting virtually all the information it can gather on its users. The company also maintains the right to purchase information on shoppers from data brokers as well as to gather information that shoppers provide retailers "in the course of using their services." shopkick further reserves the right to share any of this personal information with merchants so long as shoppers give shopkick their various merchant reward program account numbers (which many do out of convenience to simplify awards transactions), or identify themselves as a shopkick member (to earn points) when they make a purchase from a shopkick-related store. Only those customers who bother to wade through the privacy policy can learn of these conditions, as shopkick does not otherwise mention them. And although shopkick won't share shopper information among its retailer clients—for example, it won't tell Best Buy what a customer bought at Macy's—it does compile all that data for itself and maintains the right to use it to analyze shopping behaviors and target shoppers with ads accordingly.

First Data Corporation, a major credit card payment processor, has blended its primary function with managing rewards programs for smaller businesses that otherwise aren't set up to operate one.[58] Using beacon technology, merchants are notified when rewards shoppers open the First Data's Perka app upon entering a store; the notification includes the customer's name, a photo, and most recent transaction at that business. Shoppers may appreciate a personalized greeting and offers that result from their presence being announced, but they likely are unaware that all the personal information that the app gathers on them in the process is shared with many sources outside that particular store. Perka's privacy policy is even broader than shopkick's in terms of allowing outside firms to purchase access to individual shopper data. Because customers use the app at any number of associated small- and medium-size businesses, Perka can harvest enormous amounts of information about the shoppers who use the app and visit the websites of these merchants. Based on a customer's initial registration when downloading the app, and on subsequent behavioral tracking as the customer uses the app, shopper profiles are built around "personal data such as your name, birthday and email address, any transaction history, and could include a photograph." First Data mines "geographic location details" both outdoors (via GPS and Wi-Fi) and inside stores (via Bluetooth) so shoppers can be sent "relevant and appropriate offers."[59] While it shares just transaction-specific data with the merchant making the sale—and not the extensive personal information the company has compiled—First Data also sells access to third-party advertisers "to serve advertisements regarding goods and services" based on customers' activities on Perka-associated websites as well as on other websites (which are monitored via cookie-like trackers).[60]

Merchants who use Perka are generally aware that they are not in full control of their customer relationships, and this concern at least partly explains why large retailers release their own apps even as they also work with outside loyalty trackers such as shopkick, and proximity ad networks such as inMarket. Several people interviewed for this book pointed out that the premier retailer

Macy's, which employs shopkick's perimeter-and-beacon app, also fields its own app. The retailer values shopkick's ability to help bring desirable customers into the store and assist Macy's in identifying them, but at the same time it doesn't want an outside technology firm to have complete control over its customer data. Unlike shopkick's app, the Macy's app is not an explicit attempt to encourage shopper loyalty. Its dominant focus is on encouraging omnichannel shopping at Macy's either remotely or inside a Macy's store. The app invites product searches via text messages, product ID information obtained online, a smartphone-generated photo of the product, or the item's barcode scan. Neither logging in to an existing account nor registering for a new account is necessary to access these or many other of the app's functions, so any tracking that occurs under those conditions is anonymous. However, the app does encourage users to set up an account and create a personal profile. When shoppers do register the app, Macy's can match the personal information they have provided, such as name, birth date, and physical address, with information obtained about them from shopkick and other sources.

Many apps often offer features such as ease, special offers, and game-like activities, which entice shoppers to accept smartphone settings that link their device with cellular, Wi-Fi, Bluetooth, and other technologies while enabling retailers to track individuals walking toward, through, or away from their physical locations and gather additional valuable information as they do so. A few examples:

- Old Navy and some other retailers use sweepstakes offers distributed in widely broadcast anonymous text messages as a way to begin collecting personal data. Once a cell phone user opts in and provides the requested personal information, a company can "start to develop the data set you need as a marketer to actually deliver personalization," said an executive who was familiar with Old Navy's digital-marketing agency and who requested anonymity to be interviewed for this book.[61] "The consumer texts in, and there's a prompt that responds. It says, 'Are you sure you want to join? Reply yes.' And some companies will say 'Reply yes and include your birthday' as the next validation. Once the consumer has opted in, now that mobile number is in your database, and you can link that number with your CRM [customer relation management] system."

• Although Walmart doesn't have a traditional loyalty program, it has nevertheless gathered personal data on its customers by asking them to register on its website and app and by using credit card information that customers provide during a purchase. In 2015 the giant retailer introduced a new vehicle by which it could identify individual customers: under its "Savings Catcher" program, shoppers submit the prices of goods they've bought by scanning a barcode on their purchase receipt using the Walmart app (they can also input the barcode number by hand on the company's website). Walmart says it then compares the prices against those listed by its competitors in the past month and will issue a credit for any difference in the form of a Walmart gift card.[62] To participate, customers need to identify themselves, whether they either already have an account or paid cash for the transaction. Walmart can then connect this information with other purchases a customer has made. The program also encourages customers to download the Walmart app to simplify access, and those who do so can then also be tracked by beacons when they shop in the physical store.

- Target's popular app helps shoppers aggregate discounts, compare Target coupons with those of other retailers (and possibly receive lower prices as a result), find products in the aisles via an interactive map, and check an item's availability. But between the app and other ways customers interact with the retailer, according to the firm's privacy policy, Target and its "service providers" use "cookies, web beacons, and other technologies" to record just about every such action it can, and by every means—"online, in store, mobile, etc."[63] The phone app also connects to inMarket's beacon technology for further data collection.[64] Target's privacy policy for its app states that the chain buys additional information from third parties and shares this data with its subsidiaries and affiliates, as well as with other companies "for their marketing purposes."[65] Target's main privacy policy adds that "if we are able to identify you as a Target guest, we may . . . link your activity on our website to your activity in a Target store. . . . This allows us to provide you with a personalized activity regardless of how you interact with us."[66]

Target is one of the few companies that present a separate privacy policy for their app, as most use one policy to cover all aspects of their business. This variation notwithstanding, privacy policies in general often mask rather than reveal firms' handling of customers' personal data. shopkick's long, complex privacy policy is typical; it uses terms such as *non-personally identifiable information, affiliated partners, API calls,* and *psychographic data* without defining them or providing any explanation.[67] Ulta's privacy policy is likewise characteristic in that it doesn't explain how it profiles people based on the information it gathers or how the data translate into what it calls "special offers and personalized content."[68] Privacy policies also often seem contradictory when they describe their data-sharing activities. For example, in one section the Safeway supermarket chain privacy policy states, "We do not sell, rent, lease, share or disclose personal information to any non-related companies or third parties without your consent," except for certain narrow business purposes like filling orders. But in another section the policy notes that it allows "third parties to serve ads on our websites." It also says that the company shares personal data from cookies with the third parties. Marketers are well aware that the information those third parties collect in this manner enables them to easily identify the names and other personal information pertaining to Safeway customers. And because the supermarket chain charges the marketers for serving the ads, indirectly at least Safeway does "sell . . . personal information."[69] Attorneys representing various marketing organizations acknowledge privately that these policies are not written for public consumption. Rather they are arcane documents that provide the retailer with legal protection, so they mention everything the merchant does to learn about its shoppers but without revealing specifics that might enlighten competitors or stir up the general public.

Retailers continue to seek out new ways of combining data collection with benefits so that shoppers fixate on the perks and don't stop to consider the behind-the-scenes activities that capture their personal information. Mobile wallets are among the newest such vehicles. A wallet is a default smartphone app that began as a means by which users could store and access items on their device such as retail coupons, airline boarding passes, and event tickets. It has since grown to include credit and debit cards, as well as retail rewards cards. Both Apple and Google offer a version of the app. But despite the availability of mobile payments from these and other entities as diverse as Amazon, PayPal, Alipay, and a consortium of retailers that includes Walmart, most Americans have yet to give up their plastic cards and paper currency and use their mobile phones instead to make their purchases. The competition for phone-payment supremacy among these companies is likely to be drawn out and messy.[70]

Battles among mobile payment providers aside, some marketing professionals believe that mobile wallets can serve as at least a partial replacement for individual retail apps. Although a retailer's app can help the merchant by collecting customer information, encouraging their loyalty, tracking their locations, and sending them messages, many shoppers either never download the app or, when they have downloaded it, never turn it on. Mark Tack, vice president of marketing at Vibes, a mobile technology marketing company, noted that smartphone users access just five of their downloaded apps 80 percent of the time, and so retailers should change their strategy and tap into the mobile wallet platform already on the phone. "When you have that [wallet] app that is offered by the operating system, it sits on your phone, it's just there," he said. A report from the marketing consultant agency Forrester Research agreed, noting, "Marketing leaders will benefit from mobile wallets if they tie together loyalty programs, coupons, product discovery, gift cards, and promotions to create powerful and new brand experiences in the mobile moments of their customers."[71]

Here is how the process can work, according to a Vibes executive who did not want to be named:[72] a retailer can send emails or text messages with discount coupons to people who are customers or whose contact information the retailer bought from third parties. If the individuals open the messages on their smartphones and click on the coupons, they can deposit them into their phone's mobile wallet for later use. These promotional items can incorporate a unique identifier, which, when combined with the geolocation functions of the smartphone, can result in the coupon or other offer automatically appearing on the phone's lock screen as the shopper nears the store (or, with Wi-Fi or Bluetooth on, the location in the store) for which the offer is intended.[73] Vibes performs these services for its retail clients, and, depending on the data analysis of the smartphone user, it also offers the option of altering the text—and even the discount amount—of the coupon once delivered, until the offer is redeemed or deleted. For example, Men's Wearhouse (a Vibes client) can first send a coupon discounting the price of a pair of pants, then increase the discount or change the discount item to a sports jacket. The company can also increase the rewards points offered with the purchase, or present an offer to buy a gift card. (Interestingly, the hardware hasn't quite caught up with the rest of the technology, as a smartphone's barcode can't always be read by a store's scanner. Consequently, Vibes includes the numerical digits with the barcode so the clerk can key in the number by hand when this problem occurs.)[74]

This ability to change the value of a discount based on a company's knowledge of an individual across space and time raises coupon personalization to a new level. Making the activity a common promotional tactic requires retailers to make loading retailer messages into a mobile wallet part of shoppers' regular routines. That will involve encouraging them with help, special treatment, and gamelike entertainment around the smartphone. Another Vibes executive listed incentives such as loyalty points, rewards redemptions, and contests as ways to present "more opportunities for . . . consumers to save content to their mobile devices." He said that email, text messages, the internet, apps, direct mail, and QR (quick response) codes all can be used to transmit these incentives to targeted individuals.[75] And behind the scenes Vibes can tell "who has saved or deleted mobile wallet content" as well as "which [geographic] locations consumers save to their mobile wallet offers and loyalty cards."[76]

Whatever the tools that marketers and retailers use to gather information on consumers so that they can be sold to more effectively, the main goal is to make the surveillance-and-selling scenario as smooth and as tension-free as possible. The companies discussed in this chapter—Ulta, Sephora, Old Navy, Men's Wearhouse, Walmart, Target, and Safeway—each reflects the retail industry's common practice of baiting consumers to gather data on them and then to obfuscate the process by presenting confounding privacy policies. Indeed, "frictionless" and "seamless" are current industry buzzwords for a process that doesn't provoke people to stop and ask tough questions about the technologies influencing them. "By investing in mobile applications and frictionless digital payment tools that incorporate loyalty, coupons and rewards in-store," said the global head of retail consultant Accenture, "retailers can provide a seamless bridge between customers' online and offline experiences."[77] And a trade publication article on Target stressed that the retailer "plans to strengthen its data, analytics and technology capabilities to deliver more personalized digital experiences, loyalty programs and promotional offers."[78]

Such a retailing environment raises some major questions: What does personalization really mean? What should personalized messages express? In an era where personalized information can yield tailored prices as well as other offers, what should the balance be? And in what direction is retailing headed as a result of personalization?

6 PERSONALIZING THE AISLES

Mark Tack, vice president of marketing for Vibes, a mobile technology marketing company, put his point bluntly: "For us personalization isn't a nice-to-have in the mobile channel," he said. "It is absolutely required." Vibes' clients include some of the biggest retailers in the business—Home Depot, Sears, Men's Wearhouse, Gap, Old Navy, Pep Boys. In mid-2015 he estimated that 90 percent or more of the work he does for those chains centers on the physical store. He said that personalization is "definitely" a topic of discussion among his clients, and that the smartphone "has become arguably the most important marketing channel." Consumers are "living on their phones," he said, and experience a "panic attack" if they forget them at home. "Our phones actually become extensions of ourselves," he continued, and that level of personal connection means that people expect everything on their phones to relate to them individually. "Consumers are much more likely to turn off marketers from their phones than they are other channels," he said. "Consumers' tolerance for nonpersonalized messages on their phones is extremely low." In other words, he said, a retailer that wants to attract the attention of smartphone users must "personalize and tailor every mobile message to the individual."

Tack's perceptions were closely echoed by each of the twenty-one executives I interviewed on this topic. Mark Miller, strategy director for the Catapult brand-marketing consultancy, said, "Locating individuals is a Holy Grail. Speaking one to one is a Holy Grail. . . . It's all about one-to-one communication." But, he asked, "Can we always get there?" He flatly answered his own question: "No." Most of the other business leaders made the same point when describing their attempts to link technologies to data with the aim of tailoring messages that would get individuals to take certain actions near, around, or in physical stores. The words "experimental" and "testing" were mentioned often. "That stuff is really, really nascent; it's highly experimental at this point," said Gary Stokes of the Signal Digital consultancy when discussing his clients' approaches to in-store tracking. Randy Jiusto of the Outsell strategic advisory firm agreed, noting that retailers have been testing beacons in certain stores as they attempt to determine "the best combination of the right services to offer, and how it relates to their analytics." Jason Goldberg, an executive at the Razorfish marketing communications agency, noted an "appetite on the part of retailers to say, 'Let's do some tests to try out some innovative self-service experiments using the customer's phone.'" Bruce Biegel, managing partner of the Winterberry strategy consultancy, pointed to the broad uncertainty standing in the way of the personalization grail. "They are testing," he exclaimed, referring to physical retailers and their digital-marketing agencies. "They are testing everything."

To Biegel, the advantage of retailers in the current competitive environment would devolve to those who have "the most experience in CRM [customer relationship management] data management, data aggregation, campaign management, and analytics" because they can use that experience to "deploy a new technology, develop testing criteria, and roll it out faster than those that don't have the experience or the culture." He saw Macy's as a frontrunner. "I believe that Macy's, because they have the analytics bench, because they've got the database and maturity in their organization, and because they have analytics talent, they are better positioned to continuously test. And any marketer that we've worked with who's really, really good and really analytically driven, they will use anywhere from 15 to 30 percent of budget for testing. They're trying to beat that control all the time. And so a Macy's should always be testing. The whole job is to beat the control."

Razorfish's Goldberg agreed that Macy's has implemented lots of testing to encourage optimal customer in-store involvement as well as to gather data toward understanding shopper patterns and personalization. He noted that the retailer was testing a number of loyalty programs, including Plenti, a trading stamp–like loyalty plan from American Express. He said that Macy's was one of the first to enable customers to scan QR codes with their smartphones to obtain information about specific products. It was also one of the first to invest in mobile visual search, which enables a shopper to submit a photograph of an item to the Macy's mobile app to find out if that product or a similar one is available in the store. "They are doing the beacon stuff," Goldberg said. "They are doing GPS [tracking]. They are . . . partnering with shopkick. They're doing some in-store analytics . . . to understand what the customer flow is, how frequently the same customers come back, and what the conversion rates are. They have a whole host of initiatives like that."

But one executive, who asked for anonymity, was not nearly as enthusiastic. He gently scoffed at the image Macy's promotes in the trade press and at retailing industry events. "Macy's is a very good tech company," he emphasized. "They are willing to try any ideas or concepts that come into vogue. But they are far less organic about it than they would have you believe." He described the retailer's activities as fragmented and noted that even though Macy's hired "the first omnichannel officer in all of retail," in actuality it's "really an honorary position with no specific responsibilities and no real leverage within the company." He said that the retailer assigns "a pizza box full of employees—you know, six employees"—to pursue disconnected company initiatives and "launch them all off and learn from each in their own little vacuum."

This same assessment could be applied to many physical retailers in the mid-2010s. Despite a generally agreed-upon understanding of personalization and its relation to segmentation, there currently exist quite different perspectives on how best to use customized messages to reach shoppers. Moreover, the industry is struggling with key questions about what should be conveyed to shoppers via the smartphone: should the messages be mostly tailored salutations welcoming them to the store they've just entered? Should the communications mostly focus on pointing out certain products that fit a targeted customer's profile? Should they present price adjustments aimed at offering just the right discount that entices a particular shopper to buy? Should other subjects—for example, loyalty, personal hobbies, provocative food choices—be included in the messages?

One aspect seemed clear: the data customers gave up, either knowingly or unknowingly, contributed to a data fest of experimentation. These trials created profiles of individuals that, when wedded to new targeting technologies, pushed the impulse toward discrimination to high gear. Shoppers might not be able to quite understand the hidden curriculum that urged them to give up information through various forms of bait. But they would increasingly feel how it would affect their movement through the twenty-first-century marketplace.

Joe Stanhope of Signal Digital mirrored the comments of most of those interviewed when he defined personalization as a "kind of data management where we collect data from lots of online and offline sources and then bring the data together for a single identity, and then make that data available to retailers many times." For most of his clients, he added, personalization also involves creating a constant experience no matter the means a retailer uses. "So what I [as a customer] see in my laptop on my tablet and in my mobile, in the store, when I get mailings from the loyalty program there is some degree of continuity and consistency." While not disagreeing with Stanhope's broad description of personalization, Jason Goldberg of Razorfish believes that the concept is too vague to dictate specific activities that will solve specific retailing problems. He said merchants using the term are instead just reflecting panic over a new environment rather than an actual common understanding of how to carry it out. "If you interview certain retail execs," he said, "[personalization] always shows up as the number one or number two point on the road map—'Oh my God, I need more personalization online and in the store.'" But, he argued, "what that aspiration is varies widely from retailer to retailer. So for one retailer personalization is automated product recommendations, and for another it equals a 360-degree view of the customer with client-selling tools for their sales associates. And for another, personalization exclusively means customized pricing promotion. So you can look at the studies and say, 'Oh my God, personalization is hot,' but the word is so amorphous that it almost isn't a useful data point to talk about personalization in the abstract. You can talk about individual tactics, and some of them have a more demonstrable return on investment than others. But it's difficult to just talk generically about personalization."

These two descriptions reflect the complexity involved in attempting to define and then institute personalization as a technique for reaching customers persuasively. For one thing, the word has charismatic charm within the retail industry; if done well, many see it as a key to meeting retailing's marketing challenges in the twenty-first century. For another, the word is kaleidoscopic; the meanings retailing executives and industry consultants attach to it vary depending on their business needs. When it comes to enacting personalization, some executives focus on store aisles and checkout areas as the ideal locations for sending messages, while others emphasize means far broader than the physical store. Company reports, industry trade-paper discussions, and executives interviewed for this book go back and forth between these approaches, depending on the topic at hand. One aspect remains clear through it all: new norms are coming into being regarding the ways customers are viewed in the new shopping world.

The Neustar analytics consultancy tried to canonize some of those norms in a 2014 report titled *An A–Z Glossary of Personalized Marketing*. Noting that "many marketers are struggling" to accomplish "personalized digital marketing," Neustar said the business "has grown so quickly that it's developed its own vernacular." It added, "This guide has been created to give you a better understanding of the components of scaled, personalized marketing, so that you can give your customers the one-to-one dialogue they're waiting for."[1] Significantly, only four of the forty-three glossary entries mention physical retail in their descriptions. And the report has no entries for beacon, proximity (NFC) chip, lock screen, clientelling, geolocation, and a host of other terms crucial to understanding personalization in the physical retailing context. Still, Neustar made it clear it saw the physical store as a crucial part of the new personalization ecosystem.

The Janrain retailing consultancy took a similarly broad view of personalization. In 2014 and 2015 reports, the consultancy nested its understanding of the term within a model of marketing continuity and identity-driven marketing. It described marketing continuity as "a framework for thinking about and planning for an increasingly fragmented marketing landscape." Identity-driven marketing, in turn, involves creating a "comprehensive, centralized and universally-accessible customer identity that informs every marketing message or offer an individual receives."[2] This continually refreshed "identity management system should become the driver of segmented and personalized messages."[3] Janrain defined segmentation as "a way to get a 'compelling message' in front of a smaller or more targeted group of customers who share similar characteristics." It was, the consultancy said, a pragmatic waystation between the supposedly earlier "spray and pray method of marketing" and highly specific messages based on detailed knowledge of an individual.[4] Recall from Chapter 5 Janrain consultancy's proposition "Not all customers are created equal, and not all segments are, either." A report from the firm advised marketers to divide target audiences into groups based on psychographic as well as demographic information. "Beliefs, values and attitudes are far superior predictors of response, affinity and loyalty [than] the typical age, gender and location data on which many marketers rely," it asserted.[5]

As useful as discrimination through segmentation can be, Janrain said, it is actually a relatively weak form of personalization. The stronger type, "truly one-to-one marketing . . . goes deeper to enable customized experiences based on individual-level data, such as self-declared affinities and interests or past transaction history."[6] True personalization is "perhaps the most difficult identity-driven marketing tactic to put into practice."[7] It requires "real-time responsiveness and relies on highly accurate and comprehensive customer data." Because of the difficulty of collecting the identity-linked information and setting up technology capable of analyzing and applying that information to messages in real time, "Most marketing organizations have really yet to implement systems and processes that enable true one-to-one marketing experiences." Nevertheless, Janrain reported, "a majority of marketers believed [in a 2014 survey] that personalization would become the most important capability for their teams in coming years." The ones that do it early, it claimed, "will be far ahead of the competition in terms of differentiation and added value."[8]

Joe Stanhope emphasized an additional continuity between pursuing narrow segments of customers and specific individuals. "A segment," he said, "is really any kind of rule that's applied to the universe of customers." It might be based on the day's weather, or on whether the merchant had designated a shopper a gold or a platinum customer—or a lapsed one. An "awful lot" of his clients also "score" shoppers; they place them in categories that indicate their value to the store or the store's loyalty program. The distance between segmentation and personalization can sometimes be quite slim. Asked, for example, whether "scoring" the value of segments typically affects the messages his clients send customers, Stanhope slid into a perspective that sounded very much like personalization. "It *should* affect them," he began, and then added that "given what retailers know about their customers, if they have a way to authenticate a customer through a loyalty program or something like that, they should be able to get pretty granular because they know your purchase history, they know where you live, they know your interests, they've probably enhanced that data with third-party information already. They might've scored it to figure out what you're going to want next."

Andy Chu of Sears elaborated on Stanhope's points as he described his company's classification of its shoppers for its Shop Your Way loyalty program. Sears divides customers into various segments, and then sends specific customers within each segment individualized messages depending on the person's profile and value score. The initial, broadest segments primarily reflect shopping frequency and the "way they engage with us across channels." Shopping frequency is determined according to "someone who shopped with us over a year ago, someone who shopped with us in the last six months [but not recently], the frequent shoppers." Sears then further segments these "loyalty" cohorts based on the amount spent during each visit (their "average basket size") and by demographic information—income levels, ethnicity, home address, and the number of people in the household, for example.

Gathering that kind of information about a family, he noted, requires a way to identify a particular household—which the loyalty program provides when it asks for names and addresses. Sears also buys additional information from a list of companies including data provider giants Acxiom and Experian. Then, "we really mash those [pieces of information] together, if you will, to get a sense of who the customers are." As an example he presents "a Hispanic household that falls in the age group of thirty-five, has three kids, and income level of more than $75,000, they shop with us more than three times a year, and their basket size is XYZ dollars. So we have different attributes and then we look at different sources that those attributes potentially come from." Of particular interest are "high-value customer" groups within the segments. Chu calls one such subsegment Four-Plus Trippers. "Those are customers who shop with us very frequently, and within those we look at how much money these people also spend with us." Based on its statistical modeling, Sears places a certain value on one or another subsegment. Then, based on what it knows about the group's purchasing habits in terms of products and the channels it uses for retail purchases (the Web, the phone, the store), the company will approach individual customers with particular messages and offers tuned to their group. "If we determine you have an interest in the tools category and the last time you bought tools is three or four months ago, and we also saw you looking online at lawn and garden stuff, we might push you a notification" to encourage a tools purchase. The communication could be a text message or an email, he said, depending on what statistically had worked best to lead people like "you" in the segment to purchase the item.

While such increasingly narrow segments come close to one-to-one personalization, Ryan Bonificino, the marketing head for the Alex and Ani jewelry business, asserted his company actually carries out the real thing. He said his firm's one-to-one involvements are with those who already have bought from the chain in the physical store or online and whose identities the analytics team therefore often knows. The key to successful personalization, he said, involves optimizing the use of the company's data management platform. It contains information about all the firm's identified purchasers, including "every single digital touch point." He added that the firm has an average of six hundred strands of information on each person and "upwards of twenty-eight thousand data points so far on our oldest customers." That includes bits about activities on desktop computers, laptops, and mobile devices.

Identifying high-value customers is an important part of the customer analysis for future tailored interactions based on those data. "We'll look at a couple of different dimensions," he noted. "One is obviously the length of time they have been connected to us as a consumer. That could be as simple as signing up for email or as particular as purchasing a product online versus in-store. Making a new account, or maybe phoning [a] call center. All those little time stamps when they start their relationship with us. So that's more of a lifetime-value play. Then we have highest spenders, most frequent spenders, and we actually do most influential as well." Influential people are those whom the company has tagged as having not only many social contacts but also sway in relation to those connections. The company tracks shopper comments through a social listening product called Radian6, which traces millions of discussions about brands in the digital environment. Bonificino said Radian6's parent company can often help his company's email provider match email addresses with commenting individuals. This information enables Alex and Ani to link specific opinions to individuals on the firm's customer list as well as to note the discussion topics and their frequency.

To assess the influence of those customers on others, Alex and Ani turns to social sign-on data, which is obtained from people who log into various websites using their password from Facebook, Twitter, Google+, or other social media sites. The advantage is convenience, as the individuals can use the same username and password instead of having to create a sign-in for each site. However, they must typically accept (through a pop-up window) that the site will receive data about them and the "friends" they have on the social media site they use for logging in. Retailers often hire a technology firm (such as Janrain and Gigya) to help implement the sign-on as well as to collect and analyze the data coming from the social media sites. What they learn can be well worth the effort. "It will be the email, the networks of friends, things that they 'like'—and you can pull in a lot of other stuff," said Bonificino. The more people in the target person's network, the more likely that person will be deemed influential, especially regarding discussions captured by Radian6. All this information determines the influence score Alex and Ani gives its target customers and, in turn, the treatment they receive from the retailer.

Alex and Ani uses the data it can gather on an individual to tailor what the person sees on its website and app. When the firm has enough information to form a profile—usually after about a month once an individual has begun visiting the site—said Bonificino, "we start developing some logic around who you are and what you want to see. What keeps you on the site longer? What makes you share more on social, that kind of stuff?" As the retailer's profile of a customer grows, what the shopper sees online and on the app becomes increasingly personalized. For example, the firm may show different people different photos of the same product, and it may present them with different product details. Personalization is not quite as focused in the case of advertising messages online or via mobile devices. "We don't use all six hundred data points while personalizing" the ads, Bonificino said. In fact, he said, when pursuing customers for the jewelry charms it sells through partnerships with Disney, Team USA, or the National Football League, some of the personalization comes closer to segmentation. Many shoppers are unaware of these connections, and they may never even have shopped at Alex and Ani. The company has been attempting to identify people who are fans of Major League Baseball's Philadelphia Phillies and who had purchased charms from the team or might buy them in the future. Bonificino said he has access to the Phillies' records of charm purchases, and he also turns to data suppliers such as

Datalogix, eXelate, BlueKai, and MasterCard Advisors to learn what Alex and Ani customers purchase from other merchants (both online and by other means) in order to identify shoppers who have a predisposition for Phillies-related items.

Industry consultants stress that the success of personalization is limited by the quality of the data. To them this means accuracy and relevance, and they argue shoppers want the same when they receive personalized offers—hence the widespread and continual collection of data about individuals from as many sources as can be identified. Consultants consider anything less than that unacceptable. A 2014 Janrain poll, for example, found that 71 percent of U.S. consumers and 61 percent of European Union consumers say they have received offers which "clearly show they [marketers] do not know who I am." The poll also found that 51 percent of U.S. consumers and 36 percent of European Union consumers have received inconsistent information among various channels—online, apps, mobile devices, in-store—from the same brand.[9] "Less than perfectly accurate" data was the source of much of this problem, according to the company.

To avoid such issues, industry analysts advise retailers to focus on amassing just first- and second-party data. First-party data comes from the retailer itself, while second-party data is obtained directly from other retailers or publishers (Facebook or Google social log-ins, for example). In one of its reports Janrain noted that third-party data (from firms such as Datalogix and Acxiom) may not be current, as an individual's life events typically unfold so fast that it's impossible to keep pace with them.[10] Consultants suggest that regular triangulation of first-, second-, and third-party data is optimum. They say that to ensure the most accurate and actionable customer profiles, marketers should use multiple sources and compare "declared, first-party information directly from the consumer or identity provider, against purchased customer data." Along with other analytics firms, Janrain recommends continually tracking individuals across as many platforms as possible. It suggests to retailers that they should engage shoppers when they visit the merchants' websites or access their apps by posing questions, encourage shoppers to participate in a poll, and have their in-store cashiers ask customers for their email address and phone number. This "progressive profiling," Janrain promised, would yield increasingly specific information about individual customers over time.[11]

Janrain was among the analytic and consultancy firms that wrote only in passing about brick-and-mortar venues as places for collecting and using data for message personalization. Despite their omnichannel rhetoric they had become comfortable with the idea of gleaning data for personalization from purely digital areas, especially websites. With the growth of mobile devices, though, an increasing number of consultants and agencies, along with retailers, emphasized the use of data for tailoring messages in the physical store. "Offline Personalization Matters Just as Much," was the headline of an early 2015 marketing industry article. "There's been plenty of talk about personalizing online communications," the article noted. "But recent research suggests it's equally important to tailor offline customer experiences as well." A survey cited in the article found that "just 23% of client-side marketers worldwide personalized offline channels, compared with 88% who used email personalization and 44% who did so for websites. Agency professionals were even less likely to tailor offline efforts, at just 17%."[12] For eMarketer, the survey findings were "another reminder" of the need for retailers to provide a genuinely multifaceted experience. "Most final buying decisions still happen in-store," it pointed out. "Retailers who can tie all the data collected on a customer [across all channels, while the customer is in the physical store] stand a better chance at closing the deal."

Some interpreted the call for in-store personalization to mean bringing back the traditional one-to-one contact between merchant and customer. The publisher of online publication *Furniture Today,* for example, exhorted her readers not to "underestimate the power of connecting with customers within your very own four walls." Arguing to "Let Amazon be Amazon," she said local retailers should not overlook their biggest strength: the opportunity to connect with their shoppers and to remember that people are social by nature. Her suggestions: interpersonal conversations "that genuinely engage with customers about their day," and to provide exceptional customer service and to inspire shoppers by creating a visual counterpart to social sites such as Pinterest—but adding smell and sound. "Making the shopping experience inspirational is a winning combination with shoppers; many prefer to have the touchy, feely experience. Again, local retail wins."[13]

Others argued that by combining quality data with appropriate in-store technologies retailers could eventually accomplish the same goals—only more efficiently, at greater scale, and perhaps even more successfully. "Marketers Are Showing Customers

Love," read an April 2015 headline from the trade magazine *DM News*. The article said the "love" emanated from accurate, customer-relevant data leading to personalization.[14] While *DM News* wrote broadly about tracing "multi-touch customer journeys" as a source for personalization, a Forrester Research report in 2014 heralded "new key metrics that retail store analytics will unleash" for "insights that will drive increased engagement with customers as well as greater efficiency in retail store operations." The report went on to say that "retail stores have been living in the analytical 'dark ages' in comparison to digital channels," and that recently "technologies such as Wi-Fi, Bluetooth, GPS, and video" have allowed retail store analytics to gain a foothold among digital retail businesses, although "the depth of these analytic tools is lacking and immature."[15]

Forrester also noted that physical stores were beginning to view personalization in and around the physical sales arenas on par with that for the Web. Retailers' experiences with this varied; not only were they targeting different audiences, but each consultancy and marketing agency had its own philosophy and so glommed onto different technologies to carry out the targeting goals. The mobile technology company Thinknear, which approaches shoppers between home and store, hyperbolizes on its website that its technology is "so accurate, it's insane. . . . We've combined location technology, a deep knowledge of consumer behavior, and first-party data from our parent company, Telenav, to give you the ability of a mind reader—or some crazy shaman—so that you find your audience and finally deliver those mind-blowing ideas." Telenav, a wireless location-based services company, offers an app featuring maps, navigation support, points of interest information, real-time commute times, and weather forecasts. The app is free; the company makes money capturing, storing, and analyzing information based on individuals' use of the app so they can be targeted with commercial messages. According to Daniel Mahl of Thinknear, a special benefit of the app for his firm is that it continually updates latitudes and longitudes of restaurants, stores, train stations, airports, and other points of interest. Thinknear uses those locations in two ways. It can target people who are in certain areas, and it can target people who *have been* in those places.

Say a client wants to reach travelers who are near its stores. Thinknear goes after those types of people in two stages. First, Thinknear computers link to mobile-advertising exchanges that auction individuals based on their mobile phones' current latitude and longitude. Thinknear bids on locations frequented by the types of people its clients might prefer—John F. Kennedy International Airport, for example, to reach "travelers." If Thinknear wins the bid, it can serve its client's ads to the particular phones at the airport. Just as important for Thinknear at this stage are the data that the company gleans from the phone receiving the ad. If the ad lands on a mobile browser, Thinknear includes what's called a tracking pixel, which can record the Web page on which the client's message was presented. Tracking pixels have less utility on phone apps, but in these instances the company can record the personal information people provided when they registered for the app—gender, age, and home address, for example. The most important value of this targeting stage to Thinknear is that it provides the company with the unique identification numbers of the devices it targets. Possessing those numbers enables Thinknear to track the individuals via GPS, cellular triangulation, or Wi-Fi as they move across the landscape mapped by Telenav. The company then executes its second stage of reaching "travelers" via mobile devices. It goes back to the mobile exchanges and has them find the travelers' phone IDs among the phones they put up for bids. Thinknear then bids to serve ads to those phones, either for the initial retailers who reached the travelers at the airport, or for other retailers who want to reach travelers wherever they may be.

Thinknear's specialty is determining the devices' real-time locations and targeting them for clients who want to reach people in those specific areas. A typical geofence (meaning a defined geographic area) around a retail establishment, Mahl notes, is about a quarter- to a half-mile. Generally such a geofence is wider in the suburbs than in a city to account for cars being a more predominant mode of transportation in the former. The company uses the phones' location to interact with people when they are in or near a competitor of its client in the hopes of steering the people toward Thinknear's client instead. This can involve inducing the phone's owners to click a button that will take them to a landing page that, in turn, might offer a simple click to call the advertiser as well as directions for the client's store.

Mahl says this landing-page technique "works well," though not necessarily better than simply sending ads for his clients. Although the company can target men or women separately as well as both sexes together, he preferred instead to discuss Thinknear's focus on an individual's location as the pivot around which all persuasive activities turn. The company's software can continually remind the phone user how far the individual is from the advertised retailer. It also can change the message based on the weather, pollen count, or even UV index in the phone holder's location. "I'm working on a campaign right now [for sunscreen] that we're only serving ads to people when the UV index is 6-plus," Mahl said. He went on to say that Thinknear is working with a location-analytics firm called Placed to determine whether the app-directed mobile device display messages lead to sales even though his firm is not involved in reaching shoppers in the store. He also said that Thinknear wants to be able to match behavior on the Web and other media to the owners of the phones they target, but the company has yet to fully achieve that goal. "It's definitely getting better," he said. Thinknear's ability to identify individuals they can pinpoint geographically has increased, as has the firm's ability to gather data about those individuals from various sources. Yet, he added, "it's still a ways away from being the industry standard and even then it's still tough to get it down to an exact science and prove that it's completely correct."

In contrast, the Vibes agency downplays the value of delivering ads on mobile browsers and apps. "When you think of mobile messaging and mobile marketing and communication on the mobile phone," said vice president of marketing Tack, "we like to recommend [that] the three biggies are text messaging, 'push,' and mobile wallet." Text messaging needs no special technology, but federal law requires that marketers offer a user the choice to opt in to receive these messages. Consequently, Tack said, because the users choose to receive such messages, they likely pay more attention to them than they would to random email messages. Moreover, he continued, those who opt in to receive texts likely are more loyal, and therefore more valuable, to the retailer. "Push" messages and the mobile wallet help ensure that these messages do not go unnoticed, Tack said. By push he was referring to messages that pop in front of the targeted person on the phone's top screen—its lock screen—at a preset geographical location. The push notification is tied to the presence of the client's app on the phone. Many of their customers download those apps but don't use them. Push-notification technology exploits the app's presence without relying on the shopper to do anything with it. It is activated by cellular technology, Wi-Fi, or store beacons based on the shopper's location. If a shopper doesn't have the retailer's app (and many do not) there is another way to push a message to the person in a particular location: send a coupon as a text message and encourage the person to save it in the phone's wallet. By placing location tags on the coupons, marketers can instruct the wallet to send reminders about them to the device's lock screen when the person is in a particular place—a specific aisle in a specific store, for example.

Tack emphasized that these messages must be personalized if they are to be successful. "If you're sending a text message to someone and you just do a spam one-size-fits-all, you're going to have a pretty low response rate," he said. "What a lot of our more sophisticated clients do—like a Sears, a Home Depot, a Pep Boys, a Men's Wearhouse, as a few examples—is they'll integrate their mobile database with their CRM [customer relationship management] systems" and their loyalty information. The customers are then placed into segments and sent messages "that include the customer's name, category of products that they prefer to buy or they have purchased in the past—offers that have proven to be effective for that particular customer." He continued, "In the mobile marketing channel, for retailers to be effective in what they care about most, which is driving in-store foot traffic, increasing basket size, and improving loyalty, we have found that using a personalized approach just delivers significantly better results."

Shoppers who come to Vibes' clients via their smartphone's Web browser rather than the app are assigned a lower priority, as the agency believes browser use means that shoppers are not yet sure what, and from whom, they want to buy. Tack said the company prefers to try to learn about people who are already "engaged with the retailer, with the app installed, texts requested, or coupons downloaded." Such people merit personalization, he said, a task that's feasible "because now you actually got the data." His company advocates text messaging, push notification, and wallet-generated lock screen views as the preferred means by which to reach those shoppers.

inMarket has yet another take on the technologies retailers should use to pursue shoppers. The company is struggling to marry two businesses, outdoor geofencing and in-store beacon messaging. The separate operations frustrate inMarket executive Jeff Griffin. "I work with retailers, and retailers care about two things," he said. "One is driving trips to their store. Geofencing does a really good job of driving trips to a store. . . . [But] it's not very good for driving the second thing that retailers care about, which is conversion—making a sale. Because you can't penetrate the building. Beacons are the thing that allow retailers and brands to do that. And when you're in a store to buy something, you can receive that message, whatever that motivational message might be." He gave an example: "Hillshire Farm, which is a really good customer of ours, they would deliver messages to buy Hillshire Farm sausages within one mile, for instance, of every Walmart in the country. And that message, frankly, can't penetrate the inside of a Walmart. You only get it when you're in the parking lot."

To link the two businesses and expand inMarket's marketing possibilities, the company has been experimenting with targeting shoppers outdoors based on its successes with the beacons it has placed inside stores. For example, as a result of ads the company has delivered for Hillshire Farm within supermarkets that carry its beacons, the "interaction with the brand" within those stores, as well as the number of shoppers installing and using the Hillshire app, increased substantially.[16] The next logical step, Griffin said, is to note via beacons where people linger in a store, infer the products [Hillshire's or others] they might be looking at, and pursue them through geofencing with ads about the items after they've left. This practice, he contends, will give the physical retailer the same power as Web merchants to personalize ads to shoppers beyond the store's borders.

He offered the example of a man pausing to look at a lawn mower in a store. The beacon would register that he lingered for a substantial time, indicating an interest in the product. Griffin said the retailer should be able to send a message to the person about lawn mowers when he leaves the store, much as a person who viewed a mower on a website might find himself encountering lawn mower ads while browsing elsewhere on the Web. inMarket's location-specific ad business could do just that, said Griffin: "Once that person leaves the establishment and gets out of the range of our beacons, the fact is we can at that moment begin to drive retargeted messages to that phone that have to do with [the product]." Yet he acknowledged in mid-2015 that "we don't do that much yet." Because inMarket is "only in a couple of thousand stores"—and on only a dozen apps—it has been challenged to gather sufficient information on what shoppers are doing inside stores to enable it to engage in large-scale retargeting outside stores. He also said that many of his retailing and advertising associates have yet to understand outdoor retargeting: "I sit down with retailers who've been in the business twenty, thirty years, and they say, 'Get out—you can't do that!' And I say, 'Yeah, absolutely you can. It's not that hard.' "

The various challenges notwithstanding, inMarket has been able to connect at least some of its outdoor and instore activities. Using its beacon data to analyze the shopping cycles of the millions of shoppers who hold its app, the company concluded that it could determine when particular customers would be due for a return visit to a particular store and send ads to those people's phones, especially when they are near the retailers, to entice them to stop in at the retailer. In late 2015 the company reported that sending such ads on behalf of an unnamed client resulted in an 8 percent increase in store visits and a 14 percent increase in the amount of money spent per visit over the period of a year. The online media and marketing publication *MediaPost* recognized the synergy: "While most consumers may never be aware of beacons," it said, "they may find themselves receiving more ads that matter at the right time."[17]

inMarket wasn't alone in attempting to profit from sending personalized messages to shoppers as they move toward, through, and away from a physical store. By mid-2016, Swirl, another location tracking company, was working with retailers including Lord & Taylor, Urban Outfitters, and Alex and Ani to combine beacons with Wi-Fi and GPS tracking for anonymous indoor and outdoor tracking. Swirl could identify phones and follow them across different locations, drawing inferences about the users' interests and sending them messages sponsored by one or another merchant. For example, reported MediaPost, "a retailer could send a message in a parking lot when within a certain number of feet of a store offering a deal at the store. When that shopper enters the store and signs on to Wi-Fi, another location point is captured. And when that shopper nears a beacon in a department, yet another data point is captured. All that data could be used to trigger an instant, target message or offer or simply filed away for later use."[18] "Later use" is a particularly attractive feature, MediaPost noted. Although Swirl's basic tracking would be anonymous, it did tie the data to the phone ID. It would then hand the tracking information, along with linked IDs, to the retailers where the smartphone owners shopped. The retailers could match the phone IDs with personal data they had stored via loyalty programs and other means. Anonymity would dissolve, and a new trove of material to build ideas about whom to target, when, and why, would come into play.

Twitter was one of the companies investing in Swirl's new enterprise. In 2015 the social media giant was itself working hard to show that tailored ads placed alongside its social messaging could be a boon to retailers as shoppers moved through various daily experiences. In earlier efforts to increase revenues, Twitter had integrated the ability to place commercial messages carrying photos, videos, and links to advertisers' websites into the flow of a subscriber's feed. In 2014 it incorporated a Buy Now button to enable mobile users to purchase an item directly from the merchant whose product was being advertised, sometimes even using a credit card number they had stored with Twitter, without leaving the social media app. Twitter considered the development so important that it designated staff to work directly with very large retailers, including Walmart, Macy's, Best Buy, and Walgreens.

Twitter's Patrick Moorhead was assigned to work with Walmart. Part of his job is to identify people on Twitter beyond the currently active Walmart followers on the social network. "I'm probably reaching twenty million consumers today on Twitter with messages about Walmart's fresh groceries that have nothing to do with the six hundred thousand people who follow the Walmart account on Twitter," Moorhead said. Twitter's agreement with Walmart incorporates a revenue sharing model that encourages Twitter to work toward achieving sales for the retailer online, via mobile devices, and in the store. Twitter does not charge the retailer directly for ads; rather, the social messaging site receives a portion of the sales revenue either when a person purchases a Twitter-advertised product using the Twitter purchase button on the Web or the mobile app, or when a shopper who received Twitter's Walmart ad for a product subsequently purchases that product in-store (database coordination between the two operations identifies the link). Moorhead sees the program mainly as a way to reach consumers outside the physical store rather than in it (or rather than those who might travel there because of the Twitter program). "It's less about Twitter being present at the shelf at Walmart and more about creating the ability for Walmart to export the shelf to consumers wherever they are," he said. At the same time, he noted, "over 70 percent of Twitter users use Twitter while they are shopping, looking for product information, often reading the conversation about products even if they don't participate in the discussion." Consequently, he acknowledged, the Walmart ads could spark purchases in the physical store, and he noted that Twitter has the ability to use location as a consideration when it targets individuals with ads.

Moorhead believes that Twitter is uniquely positioned to reach large numbers of individuals based on their stated interests and pursuits. "We have very robust data science organization within Twitter," he said. "On Twitter, people tell us what they're interested in and they tell us what they feel all the time when they tweet. So we have built technology behind the scenes to compile all that information and operationalize it as, 'Hey, find me everyone that via their Twitter activity—you know, who they follow, what they read, and what they tweet about—[who] are highly likely to be interested in fresh, cheap groceries from Walmart, either for same-day delivery or in-store purchase.'" He said that Walmart and Twitter mesh well. "When we look at Walmart we know their strategic growth customer is the millennial mom—younger mothers with kids in a household." He explained that Twitter can analyze people's tweets to unearth all kinds of information about them, such as whom they follow and what they read. Armed with all these data, the company can then zero in on specific demographics according to personal interests. For example, he said, he can target those who indicate by their tweets that they are interested in looking for recipes, or those tweeting about healthy food for their kids.

Walmart, for its part, has a world-class customer tracking and analytics division it calls WMX. According to Moorhead, it knows which products huge numbers of specific Americans and American households put into their shopping baskets, and in which combinations. Walmart tracks its individual customer purchasing habits by connecting them via their use of credit cards at the physical store's register as well as their habits on Walmart.com and its related app. In their collaboration with WMX, Moorhead said, Twitter's software experts built an infrastructure to enable Walmart to find its customers in Twitter's vast database of registered users. An outside matching service then takes the data from WMX and matches it to Twitter users based on email addresses and other user account information. Once these connections are made, Twitter can analyze specific populations and separate them further based on its own data on the individuals and their interests before targeting them with personalized messages. For example, Walmart can identify Twitter users who are millennial moms who shop at Walmart for soy milk and gluten-free bread and who may be good prospects for a certain yogurt. Twitter takes this information and conducts further analysis on this group: Moorhead suggested hypothetically that "they all tweet about whatever product Oprah gives away each week, they all tend to be interested in yoga, they all tend to be over thirty, they all tend to be interested in fashion, and they all like

Taylor Swift." Twitter uses all this information to target Walmart ads to Walmart customers as well as to non-Walmart customers with similar characteristics (for example, Twitter users who don't shop at Walmart but who have profiles that parallel those who buy soy milk and gluten-free bread).

Soon, Moorhead went on, the Twitter-Walmart connection would extend into an additional venue: commercial television. Many people tweet about the programs they are watching, and Morehead said that Twitter can monitor that activity and correlate it with Walmart commercials airing on television. Once Twitter identifies the person tweeting about a television program during which a Walmart ad airs, that individual would likewise receive a Walmart ad on Twitter. "With the buy-now functionality on Twitter," Morehead said, "we would essentially have the ability to know that Joe [is] sitting in his living room with his phone in his lap watching CSI on CBS, and at 6:19 p.m. Joe saw a Walmart television commercial featuring Weber grill tools, and then at 6:23 p.m. we would serve Joe a promoted tweet from Walmart [inviting] him to buy those tools on Twitter." The seemingly personalized message could lead "Joe" to purchase the tools in a Walmart store—and the chain could track the transaction and pay Twitter for its involvement. "The whole scenario that I just painted for you . . . sounds like it's an episode of [the surveillance-themed movie] *Minority Report*," he exclaimed enthusiastically, "but that will be a reality by the end of November [2015]."

None of Walmart's messages offers a discount. The merchant, Moorhead says, instead wants to underscore that it offers low prices as a matter of course. ("Rollbacks," which may seem like discounts, are not the same thing: Walmart offers them when it persuades a supplier to provide a product at a lower wholesale price so the retailer can pass the reduced cost on to the shopper.) For many other retailers discounts are a part of life even though they're not necessarily profitable. Executives acknowledge that they have been feeding Americans' virtual addiction to the appearance of a "sale" for decades through a variety of means—especially coupons in the grocery, drug, and chain store arenas. Even brand manufacturers as powerful as Procter & Gamble haven't been able to break the habit. The mobile phone hasn't changed the situation, though it has introduced new challenges, such as the need to calibrate the number of messages, and the mix of different message types, that retailers and their consultants send to shoppers' phones inside and outside the physical store. In addition, when retailers choose to use price as an incentive, they must decide whether it should be personalized or not—and if so, how.

Jason Goldberg of Razorfish is among those who urge caution regarding discounts in the digital mobile world. "I am not a huge fan of the ever-increasing ways to deliver discounts and promotions," Goldberg said. "It's mostly a race to the bottom; we've mostly taken all the margins out of the stores already. The only way that I can come up with a sustainable win is to develop unique shopping experiences that the customers value beyond price. And so I'm much more interested in figuring out ways to use this technology to figure out a differentiated shopping experience that customers actually appreciate and will seek out than I am just using it as another delivery vehicle for promotion." Mark Miller, an executive vice president at Catapult Marketing, agreed. "I'd rather not lead with price," he said, arguing instead for the traditional formula to convert shoppers to a brand: a great product, great service related to the product, and a competitive price that together ignite great word of mouth. Miller added that in the new digital retailing arena, helpful "content" or other "enriched" experiences connected to the product and its use is a stand-in for great service. One such example would be a clothing store's app that offers ensemble suggestions from different departments based on the customer's shopping history and an initial choice of items.

Abhi Dhar, Walgreen's chief technology officer, noted that Walgreens does not guarantee the lowest prices, nor does it match prices offered by other stores. Instead, his company's marketing efforts are aimed at getting customers to focus on the chain's service message—convenience and a positive shopping experience—as opposed to prices. Personalized messages tailored to a person's shopping interests and background can reinforce this objective, Dhar said. Miller says his company offers a more quantified way to take a shopper's mind off price: an armamentarium of customer-analysis tools that Catapult has assembled for such clients as Mars Incorporated, Frontier Airlines, and Diaggio, so that it can decide whom to target and which messages to target them with. In particular he pointed to Catapult's access to corporate parent Epsilon's "massive database" of people's purchases—"100 percent household penetration of data with thousands of attributes"—that provides demographic and psychographic traits along with individuals' shopping journeys to enable Catapult to infer "some of those micro needles in a haystack, insights that will drive consumer conversions toward a brand."

For example, Miller said that to better understand different types of feline owners on behalf of Mars (which manufactures cat food in addition to candy and many other products), Catapult analyzes customers' entire shopping basket when they are buying cat food. He also pointed to the firm's additional capabilities resulting from Epsilon's purchase in 2014 of the digital-marketing agency Conversant, which continually buys data relating to the transactions of millions of individuals at more than four thousand retailers.[19] The company contends these people are "anonymized" because it doesn't record their names or addresses. Yet this anonymity may be irrelevant because Conversant has so much information about each person—"7000+ dimensions," the company asserts—that by using statistical techniques the company claims 96 percent accuracy in identifying an individual's presence "across display, video and mobile." Conversant assigns every shopper a unique ID that is the basis for tracking "how they spend their time and money—online, in-store, or both." The company claims that "nearly everyone in the US" has been assigned an ID, which remain in place for "a lifetime so your conversation with each consumer never misses a beat."[20] Catapult's Miller noted that using Conversant to match his clients' offline customers with their digital presence and follow them puts his work at the leading edge of "micro-targeting." These state-of-the-art capabilities (presumably along with services from Epsilon's Agility/Harmony email targeting company)[21] enable him to answer the ultimate personalization question for his clients: "How do I deliver the right message to the right person across devices at the right time?"

Miller recognizes that retailers and brand manufacturers offer people coupons based on their characteristics and in-store behavior. He sees his role, however, as helping his clients maintain their profit margins by working with them to craft messages that would be so persuasive at the individual level that discount coupons become unnecessary. He says he tells his clients they shouldn't react to shoppers leaving their website without purchasing by automatically sending coupons to them in the hopes that they'll change their minds and buy something. "That's not good business," he insists. His goal instead is to help clients "crack that [discount] addiction" by arming them with knowledge regarding types of shoppers, and even individual shoppers, so that they can steer customers toward those "two top levers" that encourage purchasing: a great product and great customer service, "without diluting price."

In contrast, Sears goes with the addiction flow. Andy Chu, who oversees mobile management for the retailer, explained that the company carries out strong price-promotion efforts aimed at the company's Shop Your Way loyalty program members, who constitute approximately three-quarters of all Sears shoppers. Price promotions are often designed so that a shopper's overall purchases at Sears add up to an acceptable profit margin from that individual. Both virtual and physical transactions are factored into this calculation. A major concern for Chu is that some Sears customers targeted for discounts visit the store only to buy sale items—a practice that is destructive to profit margins. "As an industry, I would say we do not do a good enough job to identify who are the good and the bad customers," he said.

In the meantime, Sears continues its general efforts to identify the good customers and incentivize them to be better ones. "On a regular basis we send email [and text messages] to our customer segments," Chu stated. "We also send them loyalty points and rewards points. And all that information is based on our internal modeling. So we know that if you have affinity for the tools category and the last time you bought something is three or four months ago, and we also see that online you are looking at lawn and garden stuff—we might push you a notification [that states,] 'Hey, here are some surprise points. . . . Come spend XYZ dollars.' " He added that the offer can be geared toward fairly specific discounts, such as $5 off on apparel or items purchased in the lawn and garden department, or to more general ones, such as $10 off for spending $75 or more. "It really depends on the different modeling that we do," he said.

Chu went on to say that loyalty members don't get individual pricing, but they do receive discount offers via their mobile devices, adding that members may receive personalized reinforcement "at the points level. We might give a person additional points—effectively it's a discount. You might get $5 and I might get $10." As of 2015 these blandishments were showing up on the mobile browser and in the Sears app, but not in the physical store. Still, things are changing. For one, Chu said, prices in different Sears stores as well as online are adjusted constantly based on Sears' price monitoring of its competition. For another, Sears has also been testing "e-ink" paper-like electronic shelf-price displays, which the retailer can change instantly depending on various competitive exigencies (Chu noted that Kroger and Whole Foods supermarkets have also been moving in this direction). Still another change is evident in salespeople's interactions with shoppers in the physical aisles as a result of the retailer's digital personalization efforts. "As an example," he noted, "in the appliance area our associates . . . have been carrying iPads for about two years now. And when you are interested in a refrigerator, they will walk you through, they will show you on the iPad all the different specs, they will help you compare refrigerators. And you might say, 'Hey, I've just been looking at Best Buy. . . . This product is similar, but I can get it for X price.' And it's up to the salesperson to say, 'You know what, we can cut you a deal.' If you identify yourself as a member, they can actually look at your level—how many points you have, all that information. . . . We have something called member pricing. And if you're a member you get special discount." Chu said that Sears isn't yet able to arm its salespeople with loyalty member value-level information so they can offer individualized pricing but added that "we are starting to develop that capability."

Chu acknowledged that price incentives could be wasteful if presented to customers who entered a store already having decided to purchase a particular item. The dilemma, he said, was determining shoppers' exact situations so that Sears would know whether incentives were advisable and, if so, what those incentives should be. "All retailers struggle with this, including Sears," he noted. "Are you leaving money on the table if someone's willing to buy at the [posted] price? You don't [know]. That's the biggest challenge right now."

Chu's approach notwithstanding, most of the executives interviewed believe that merchants should focus on nonprice messages when communicating to shoppers by smartphone, at least for the near future. Mark Tack of Vibes stated that some of his customers, such as Saks Fifth Avenue, do not send out any mobile coupons. And yet, he said, as frequently as any of Vibes' clients Saks engages in mobile messaging as well as notifications through the mobile wallet. Instead of discounts, Saks sends its customers "exclusive, insider only information," such as alerts about invitation-only events. Wintergreen consultant Bruce Biegel mirrored the prevailing view within the industry, stating that companies are trying to find their way with mobile messaging in the aisles, often without making discounts stand out. He was optimistic. "I think we're going to do a better job of engagement, and engagement doesn't just have to be price," he said. "It could be product, it could be loyalty, it could be content, it could be an event, it could be sampling." Sampling occurs when a store arranges with brand manufacturers to distribute free versions of their products, such as pizza slice samples in a supermarket, to attract customers. Biegel said that databases and in-store messaging technologies should be able to identify shoppers who might want to sample the item so that a phone message or a store clerk can invite them individually to do so. Further, he noted, merchants should position all personalized communication as a reward, an indication of privilege. That reward, he said, could be many things besides a discount. But whatever the perk, "the experience should be personalized. . . . You'll come in and they will recognize you and they will say 'Hey, based on [your] prior [purchases] this is what's going on today. We are out of this but did you try that?' "

Biegel went on to say that in-store personalization efforts are constantly evolving. "There's just all sorts of crazy stuff going on," he said. Jason Goldberg of Razorfish pointed to Walgreens' in-store partnership with Google to offer augmented-reality maps—video screens installed on shopping carts that present information about products as shoppers walk by them—for the chain's loyalty program members along with messages based on their loyalty program data. Although determining the best mix of discount and nondiscount communications might be part of the motivation for the initiative, he said, "I honestly think that at this point it's more about testing the mapping and augmented-reality technologies than it is the actual offers that they deliver through those technologies." By contrast, Mars Advertising's Ethan Goodman said his clients were in fact trying to learn "the right level to which you can do personalization with some of these in-store location-based technologies" such as beacons. He wants to be able to help clients with products in the frozen food aisle to "confidently deliver a frozen food–specific message" to shoppers going down that aisle. His company has yet to find the right content that will cement the sale. "Is it a telling of how the product works, is it testimonials, it is a video, is it [a discount] offer? That's largely going to depend on the category; it's going to depend on the shopper; it's going to depend on the retailer. I don't think anybody knows [more than that] yet."

The executives also believe that shoppers need to get used to the new technologies and data-collection regimes associated with personalization. Jeff Griffin of inMarket takes a particularly conservative position in this regard, saying that his company recommends to retailers that they start by installing just one or two beacons. Of course, he said, merchants with large stores and those with more than one level, such as Macy's, should place an additional number of units, but he disagreed with Macy's decision to place beacons in multiple store departments. "We don't believe shoppers are ready for that level of bombardment," he said, noting that inMarket's analysis concluded that people who receive only one message per store visit will use the app linked to the beacon more often and keep it longer than if they receive two or more messages per visit. That's "a very bad thing," he said, because retailers want shoppers to use their apps, though he quickly predicted that this low-messaging rule would not be permanent. "Long-term, we will be delivering, and frankly [will] recommend, [in-store messaging] as an incredibly powerful advertising tool" that retailers will use to deliver both personalized discounts as well as tailored information about the store and brands in the aisles. "But early on," he said, "we absolutely believe that the shopper has to be handled with care" in order to create acceptance for the new activity.

Andy Chu of Sears was likewise reluctant to proceed with a different endeavor: sending messages (including discounts) to shoppers that Sears identifies as having inspected a competitor's product, in either the merchant's physical store or on the Web or a mobile device. "In the mobile site we have the capability to geofence against competitors," Chu said. "So if you go to Home Depot three times and then you come online [to Sears and] look at a bunch of refrigerators, we might say, 'Hey, you might go to Home Depot. We should go and give you an offer.' " To help identify these connections, Sears can purchase data pertaining to what their Web visitors view on competitor sites. Sears may also analyze its own data on people's activities in either a brick-and-mortar Sears or on the company's website and make assumptions about what they might look at when they visit competitors' physical locations. To identify when those visits are made, Sears can triangulate the location of a shopper's smartphone when the person is determined to be near a competitor store. This technique is not infallible, he acknowledged, but added that third-party marketing firms such as xAd and Place IQ have the technology to pursue it.

Despite its potential, Chu said that reasons such as "the creepy factor" are holding back this kind of preemptive personalization, as the ads might reveal to shoppers a level of tracking that they find distasteful. Moreover, some shoppers could conclude that to receive a discount offer from Sears all they have to do is shop at Home Depot, a way of thinking that Chu doesn't want to encourage. He also was uncertain about the optimum moment for delivering such messages. "Where in the purchasing decision do you want to inject those types of trigger points? Do we just send it to everybody? How do we know that the customer will be receptive to these quote-unquote coupons versus other things we could give? It's a slippery slope. A lot of retailers like us—I'm sure Home Depot and Walmart and everyone else—have been thinking about these notions." Despite the uncertainties, he said, Sears was going to proceed with pilot studies.

Chu, along with virtually everyone else associated with retail marketing, believes that shoppers will ultimately move beyond the "creepy factor." He noted that consumers who have been surveyed agree that "privacy is a big concern. In reality, what they do with their privacy setting in Facebook or in Google or everything else, I bet you a very small percentage of people go in and change the settings. So it's just a matter of time. The younger demographics already know that everything they do online is tracked. So to them it's just business as usual."

The same overall attitude will eventually extend to personalized prices, all those interviewed agreed, even if shoppers still don't like the practice. Razorfish's Goldberg pointed out that Safeway supermarket's mobile Just for You app already does "a little bit of that. You earn discounts on specific products based on your shopping behavior and so the pricing can seem completely unique to you." The store encourages every loyalty program member to use their mobile device to aim the app screen at special store displays, whereupon the screen then displays an individual member's pricing for particular products. He said the amounts are based on a shopper's previous purchases, with more lucrative offers tendered to customers considered to have a higher lifetime value to the retailer. A mobile device is a particularly valuable tool in carrying out such discrimination. Goldberg noted that tailored discounts appearing on mobile devices are less likely to cause shopper pushback than if they were displayed more publicly, such as on e-ink shelves or video screens. Ethan Goodman of Mars Advertising agreed that "doing one-to-one dynamic pricing is certainly possible." For most retailers "it probably requires a little bit more sophistication than is available now. But I don't doubt that it could happen." One executive, who requested anonymity, said his firm was already engaged in the practice and was offering customers whom its data suggested were in danger of abandoning the store higher discounts than those presented to reliably loyal customers.

In fact, some of those interviewed predicted that shoppers may enter certain businesses with the expectation that their phones will continually light up with discounts as they move through the store. To forestall unwanted bombardments, suggested Randy Jiusto of the Outsell strategic advisory firm, people may choose to turn on and off apps depending on the stores and the product brands they want to interact with while shopping. Such a situation may itself challenge retailers and brands to offer shoppers certain levels of privilege for simply keeping their apps turned on. In any event, stores will need to find ways to respond to desirable shoppers who, armed with product data, comparative prices, and a strong sense of their worth to certain brands, will demand certain prices as they stand in the aisles in front of products they want to buy. "I do think we'll have more" in the way of personalized discounts, said Signal Digital's Joe Stanhope. He predicted a range of discounts: one day some people will receive 10 percent off while others get 20 percent off on a particular product in a particular store; on another day some will see $10 off, while others are offered a 10 percent discount. Decisions on individual discounts will involve direct knowledge about each shopper as well as information obtained from look-alike modeling—arriving at conclusions about individuals whose backgrounds and behaviors suggest similarities to customers who have already purchased an item at a particular price. Marketers will also show similar types of people different prices and then use various statistical procedures to determine the optimal discriminatory approach. The need to rethink discounts through the lens of personalization is crucial, said Stanhope. "So many brands have trained their customers to shop [for] sales. That may not be sustainable, so [marketers] will start trying to get smarter to get out of that hole. Retraining customers to look for certain kinds of promotions right then [in the aisle], and to find out which customers really need those promotions and which don't. You don't have to give someone a promotion if they don't need it yet. If they are already a loyal customer, why would you charge them less?"

This approach—awarding less for loyalty—flips the traditional understanding of loyalty on its head. Yet it does make sense in a shopping world where loyalty is a vehicle for data-driven, personalized discrimination. All the industry voices in this chapter strive to find the best ways to act on this realization in the hypercompetitive and multipronged retailing environment. At the same time, they acknowledge that what they are doing marks only the beginning; they see many more transformative changes to come.

7 WHAT NOW?

"The Truth will be present in everything. You'll know everything about yourself and your loved ones if you opt in," said Jeff Malmad, managing director of mobile devices at Mindshare media and marketing services agency, which is owned by the huge WPP international conglomerate.[1] He believes that marketers should use this pitch to convince consumers of the benefits they receive from "wearables"—clothing and accessories incorporating computer and advanced electronic technology.[2] The digital watch and the fitness bracelet stand out as the most common examples today, but analysts predict this list will expand as the public adopts chip-laden glasses, headgear, shirts, coats, rings, and more in the years to come.[3]

Some retailing experts are skeptical that wearables will make a major splash. They point out that Google stopped production of its Google Glass spectacles after they were criticized for interfering with people's daily activities and for uncivilly accessing real-time data about individuals. Fitness monitors, among the first entrants into wearables (around 2014), have been accused of not retaining the attention of their owners. The consultancy firm Endeavor Partners estimated in 2015 that about a third of these trackers are abandoned after six months. Health care investment fund Rock Health noted that only half of nearly twenty million registered users of Fitbit, by far the biggest seller of fitness monitors, were still active as of the first quarter of 2015.[4]

Supporters of wearables believe that smart watches are a superior test of the success of wearables, noting that these watches can perform the functions of fitness monitors as well as other useful activities. The most prominent smart watch, the Apple Watch, was introduced in April 2015. After three months an estimated 1.9 million watches were sold. In June of that year conventional timepiece sales in the United States fell more than they had in seven years, indicating that Apple Inc.'s watch was eroding demand for traditional watches and clocks.[5] Elmar Mock, one of the inventors of the Swatch watch, predicted a month before the Apple debut that its smart watch might cause an "ice age" for the four-century-old timepiece industry. To compete, Swatch itself released its own smart watch later in the year. "The Apple Watch is going to gain a significant amount of penetration," was the more measured prediction of an executive from the NPD market-tracking firm in August 2015.[6]

Marketing analysts have taken the long view of the nascent business. "Just as tablets faced skepticism in their early days, with consumers and critics questioning the need for new devices, so too does wearable technology," wrote the international accounting firm PwC (formerly known as Pricewaterhouse Coopers) in a report heralding the growing importance of wearable devices.[7] "Wearable technology is still in its infancy," agreed the Certona consultancy in 2014.[8] The company suggested the devices will be data generators that can "improve personalization . . . by implementing predictive analytics tools."[9] For example, mileage data on a fitness tracker can alert a retailer that the customer is in need of new running shoes. But Certona also cautioned, "Google Glass, smartwatches, and fitness trackers are getting all the attention now, but brands should not discount other wearable technologies."[10]

PwC went on to note that in 2014 20 percent of U.S. residents owned some form of wearable device, either a "primary" device (a central connector for all kinds of devices and information) or a "secondary" device (one that captures specific actions or measurements that are then funneled back to a primary device). Smart watches are presently considered secondary devices, but PwC predicted that they, along with smart glasses, "will emerge as key primary devices, acting as a central collection portal for different wearables."[11] The PwC report didn't count the smartphone as a wearable, though one can argue it is a de facto primary wearable device, as it serves as a central hub for secondary accessories such as fitness trackers. If the report had included it, the percentage of Americans carrying technology close to their bodies in late 2014 would have exceeded 71 percent—in other words, the proportion of smartphone-owning citizens.[12]

The company in fact found that many of the consumers interviewed for its report think of their smartphone as a wearable device. Why, then, would anyone advocate for any other wearable? For marketers, the answer has to do with physical intimacy. The closer the instrument is to the body, the less likely people will remove it, meaning it can provide marketers with more information. A retailer app that has been installed on a person's wearable device enables the merchant to pursue a continuous relationship with the person "at home, on the go, and in store." This means an interaction with the entire multifaceted retailing ecosystem in the same way that the smartphone does, from website cookies to GPS location trackers to beacons, and much more. But because the devices are connected to the body more directly, they will have a more consistent presence than that of smartphones. Consequently, wearable devices will "usher in a new level of hyper-interconnected retail in which retailers 'join the dots' between an individual's pre-store and in-store behaviors to deliver an enhanced customer shopping experience," reported an article in *Information Age*.[13] In other words, the retailer will not have to rely on the shopper to move smartphones, tablets, or PCs "from couch to the shelf," noted PwC.[14] Instead, the intimate, always-connected nature of the wearable device will virtually guarantee continuous tracking across time and space.

Champions of wearables believe that tracking via these always-on products will be far more advanced in comparison with current smartphones. They will inevitably be part of what has come to be known as "the internet of things"—an environment in which all such devices (for example, smartphones, clothing sensors, and smart watches) connect with each other to perform services for their owner as established by the retailer or marketer that created the service software. The information generated will travel instantly to a data cloud to be run in real time through complex predictive analytics, incorporating other already-stored information pertaining to the individual. The retailer or marketer can then detect changes in the individual's habits and behaviors as well as in the individual's surroundings and target them accordingly. The PwC report notes that "this process will be made possible through passive listening elements [on the wearable] as well as active cues—what you listen to, what you 'like' and what you browse."[15] In this vision of the future, wearable devices will enable retailers to recognize individual shoppers immediately, so advertisements, offers, and loyalty programs will be individually tailored more precisely than ever. And wearables will have built-in payment systems and verifications that make checkout a breeze.

In addition, many wearables advocates say, the sustained close proximity of the devices to the users will add new elements of data to shoppers' profiles exceeding the capabilities of smartphones. PwC notes, for example, that nearly 30 percent of cell phone users turn off their phone overnight. And many of those who do keep their phones on while they sleep place them near but not actually in bed. Primary wearable devices that remain on the body both day and night transmit torrents of information on everything from sleep patterns to sports activities. The Certona consultancy notes that "nanotechnology and biometric technology is [sic] paving the way for smart clothing and turning heads in the athletic apparel industry. The data available through 'smart' fabrics is unbelievable—perspiration data, heart rate patterns, activity tracking and calorie monitoring. An innovative company can benefit from this data by pushing out inspiring messages to a smartwatch or Google Glass to keep users motivated toward their goals."[16] Indeed, Google's head of retail says that wearables "give [merchants] niche data that allow [them] to speak to customers better than ever before."[17] PwC examines some of these possibilities: "Through wearable technology, brands could present relevant content to a shopper while they are considering a product—say, in a grocery store, recognizing items a consumer has placed in the grocery cart and serving up relevant recipes through augmented reality. Brands could even tap body cues to tailor messaging. Sensor revealing that you're thirsty? Here's a coupon for smart water. Low on vitamins? Flash this for $1 off your favorite vitamin-loaded juice product. Serotonin levels down? Grab yourself a free soda and open happiness."[18]

Although the retailing establishment seems reasonably certain that people can be persuaded to allow retailers to identify them through certain wearable devices, it's not nearly so confident that shoppers will consent to participating in the other emerging retailing tracker, the facial recognition system. This technology involves taking complex measurements of facial images and converting them into a mathematic calculation called a "faceprint," which is then compared against a faceprint database of photographs and video still images. In 2014 the *New York Times* reported that "if security cameras record someone at, say, a store or a casino, the system can compare the faceprint of that live image to those in the database, taking only a few seconds to run through millions of faceprints and find a match."[19]

The various companies working with facial recognition systems use different techniques, and there is a lot of competition among those trying to make money on their face-matching strategies. Law enforcement agencies logically are major customers, as facial recognition can assist in matching video images taken of criminal suspects with mug shots. Some airports use the technology to speed security checks for employees and frequent fliers.[20] On the internet, Facebook is among the firms offering face-matching software to suggest the names of people in posted photographs.

Retailers, too, have begun adopting the technology. As early as 2009 an online digital news and information publisher commented that such systems represent "perhaps the most exciting innovation" in analyzing shoppers. It listed companies that were offering "sophisticated hardware and software suites that use inexpensive cameras mounted on screens to recognize human faces." These systems "keep detailed logs on who looks at what, for how long and when."[21] Fast-forward to 2014, when a reporter interested in "the future of shopping" noted that video surveillance and analytics tools "were everywhere" at that year's National Retail Federation trade conference.[22] In addition to serving as a potential marketing tool, retailers are interested in this technology for security reasons—for example, identifying convicted shoplifters that enter their store and notifying security personnel.[23]

A logical step beyond using surveillance equipment to monitor shoplifting is for retailers to apply the facial recognition technology to marketing oriented questions about the kinds of people who enter a store. In fact, Face-Six has sold software that enables cameras at dozens of malls in the United States and around the world to count and categorize shoppers based on age, gender, and race. In addition to these mall demographics surveys,

Face-Six can display target advertising based on facial analyses, CEO Moshe Greenshpan noted.[24] Similarly, food manufacturer Mondelez International was reported to be working on a system in late 2014 that combined facial recognition with digital shelf displays; a digital message board accompanying the display would tailor messages based on the demographic information obtained by the system.[25] For example, at an Oreo cookie display a person whom the system identifies as a thirty-something female could see a message stating that one of the Oreo Thin cookies in the package she is holding has just thirty-five calories, while a teenage boy might see a discount voucher in the form of a barcode scan or QR code for his mobile device.

Because facial recognition is passive, it has an advantage over mobile phone trackers. Using beacons to determine a shopper's age, gender, and race at the very least requires that a shopper possess a smartphone, with its Bluetooth turned on and the pertinent apps loaded. In the course of this exchange the phone owner's identity will often be revealed, so beacons do bring a lot more information to the aisle than just basic demographics. But supporters of facial recognition systems say that anonymous demographics are only the beginning of what they can deliver on the retail level. For example, several companies say they can offer retailers the ability to detect the current emotions of the people walking through their aisles. One such company's software "extracts at least 90,000 data points from each frame, everything from abstract patterns of light to tiny muscular movements, which are sorted by emotional categories, such as anger, disgust, joy, surprise or boredom," reported the *Wall Street Journal*.[26] The facial recognition company Affectiva told the

Journal that it has measured seven billion emotional reactions from nearly two and a half million faces in eighty countries. Many of the algorithms incorporate the seminal yet controversial work of psychologist Paul Ekman, a pioneer in the study of emotion awareness.

Another such company, Emotient, says that its software algorithms are based on analyses of an ethnically diverse group of hundreds of thousands of people participating in research for its clients via video chat. These analyses look for miniscule movements in the face. On its website in 2015 Emotient stated that "major retailers, brands, and retail providers can use Emotient's technology to assess customer sentiment analysis at point of sale, point of entry, or in front of the shelf."[27] The company's co-founder and chief scientist, Marian Bartlett, said that her company aims "to measure emotion and tell the store managers that someone is confused in aisle 12." Emotient's other co-founder, Javier Movellan, stressed anonymity. "We do not want to recognize who is watching. All we care about is what they are watching and how they feel about it." The firm's technology appealed to Apple, which purchased Emotient in 2016 with the aim, some suggested, of using it in a new version of its phone helper Siri to understand its smartphone owners better.[28] Like Apple, retailers are beginning to see the utility of this form of data. The *Wall Street Journal* reported that one retailer "is starting to test software embedded in security cameras that can scan people's faces and divine their emotions as they walk in and out of its stores."[29] *The Hill* noted that digital monitors in stores use small embedded cameras to detect age and gender; the ads that then appear on the monitors will be tailored to the camera's analysis of the shopper.[30] "There is anecdotal evidence that the results can be quite extraordinary," said Tony Stockil, chief executive of retail strategy consultancy Javelin Group.[31]

Of course, these systems mean that customers will be substantially more visible than they are in traditional foot-traffic videos, in which faces often are blurred and images are destroyed promptly to protect customer anonymity. And although Emotient and ShopperTrak (discussed in Chapter 4) stress the anonymity of their subjects, increasing competition is fostering a movement toward identifying shopper faces. Two facial recognition companies, NEC and FaceFirst, have developed systems in which cameras scan shoppers entering a store and identify existing or potential customers deemed important to the business. NEC says its system, "VIP Identification," is "ideally suited to hospitality environments or businesses where there is a need to identify the presence of important visitors, whether expected or unannounced." The facial matching software "can take less than a second" to identify a shopper and alert store personnel "so that the necessary action can be taken to greet or prepare for the arrival of the VIP."[32] The FaceFirst website says its system enables retailers not only to "spot and stop the bad guys," but to "recognize the good guys and treat them better." At a cost as low as $15 per month per store, the retailer loads "existing photos of . . . your best customers into [the FaceFirst database]. Instantly, when a person in [the database] steps into one of your stores, you are sent an email, text, or SMS alert that includes their picture and all biographical information of the known individual so you can take immediate and appropriate action."[33] FaceFirst CEO Joel Rosenkrantz told the BBC that "someone could approach you and give you a cappuccino when you arrive, and then show you the things they think you will be interested in buying."[34] The FaceFirst website exhorts retailers to "build a database of good customers, recognize them when they come through the door, and make them feel more welcome." To compile a database of customer photos, Rosenkrantz presented one possibility: "If a particular brand has 10,000 likes on Facebook, you could use the profile pictures of all the people who have liked it. . . . You can tell customers that if they agree to enroll [their face] with their camera, then they will be offered a discount coupon when they walk into the store, or get them to tick a box saying they agree that their picture can be used when they log on with Facebook."[35]

Rosenkrantz's suggestion to ask permission for shoppers' facial profile data by bribing them with discounts or exploiting their Facebook loyalties is consistent with the ways merchants encourage shoppers to give them personally identifiable data to be used in other tracking technologies—for example, Ulta Beauty's loyalty program, which gathers information that store associates can use as, tablets in hand, they greet shoppers entering the store (see Chapter 5). When it comes to obtaining an individual's permission—"opting in"—NEC's website is rather evasive. It mentions "matching images against what is likely to be an opt-in database of individuals they deem as important." In 2010 the retailing trade association Point of Purchase Advertising International published its "Recommended Code of Conduct for Consumer Tracking Methods," which suggested that consumers should be allowed to opt in before their personal data are collected. In ranking various consumer tracking methods, the code listed as "high risk" activities "any camera based OTD [observed tracking data] system" as well as "any method used to personally or uniquely identify consumers, when combined with loyalty program data, or third-party marketing data." It added, "While the federal government has recognized dangers in the realm of mobile marketing and healthcare and has subsequently passed laws to protect consumers, no such laws exist for data collection in retail settings."[36]

As of 2016, laws in this area were still lacking despite the increasing use of facial recognition technology. In fact, there were no federal laws that expressly regulated facial recognition software. (Only two states, Illinois and Texas, had laws regulating the collection and use of biometric data. Illinois specifically required firms to get permission before collecting and retaining a "scan of . . . face geometry.")[37] In 2012 Mark Eichorn of the Federal Trade Commission's Division of Privacy and Identity Protection stated that the FTC "would be very concerned about the use of cameras to identify previously anonymous people."[38] A 2014 attempt to nail down norms for the facial recognition activities of U.S. businesses was unsuccessful. The initiative was part of a larger project, encouraged by the Obama administration, to address consumer privacy and brought together nine privacy groups, leading industry representatives (including from Facebook), and retailer associations to negotiate a voluntary code of conduct. Coordinated by the Commerce Department's National Telecommunication and Information Administration (NTIA), the effort was doomed because the positions of those involved remained far apart. Chris Calabrase of the Center for Democracy and Technology was among those who argued that those using facial recognition should seek the consent of people before scanning their faces. He told CBS News that "face recognition allows secret tracking so any time you're in public whether you are attending a protest rally or visiting your doctor or entering a church or a bar it could allow you to be identified and your movements tracked." He emphasized that "the individual must be able to choose. If they can't, lots of entities, whether they are companies or governments relying on company databases, can use this technology to spy on people. They will never know, and they will never be able to control it."[39]

On the other side, facial recognition supporters contended that the demand for consent made little sense when the software is used to track criminals. The privacy groups offered a "narrow exemption" for security uses, but the industry turned it down, Calabrese said. Carl Szabo of NetChoice, a digital commerce trade association, said he and his industry colleagues would approve self-regulation requiring a merchant to post signs when using facial recognition technology. "If it turns out consumers love it, they will embrace it. If they hate it, they will walk away and that store will stop doing it." He acknowledged "the feeling that people might get that someone is spying on them and invading their privacy. But . . . we haven't encountered any misuse or abuse of the data. Nobody is selling the data to third parties." He added, "I don't think we are there yet for any calls for regulation. Theoretical fears make for bad laws. We also don't want to strangle this new technology."[40]

After a year and a half of negotiations, the retailers remained staunchly against the idea of seeking consent to use facial recognition.[41] Frustrated, privacy advocates abandoned their efforts, even as NTIA representatives urged them to continue. About two months later the General Accountability Office (GAO) released a troubling report on facial recognition technologies. It noted that it could find no data pertaining to how much these systems were being used by American businesses.[42] And it sided with those who worried that the activities had the potential to cause lasting harm to Americans. "The privacy issues stakeholders have raised about facial recognition and other biometric technologies serve as yet another example of the need to adapt federal privacy law to reflect new technologies," the report stated.[43] It was a remarkably straightforward call for regulation at a time when the industry was digging in against any effort to do so. In the hypercompetitive physical store environment, merchants wanted to create tracking systems as effective—or even more effective—as those on the internet.

Australia-based retailing futurist Chris Riddell took many of these developments to their logical extension in a 2015 blogpost titled "The Future of Retail—Emotional Analytics":

> Imagine for a moment that, by using a combination of Bluetooth proximity and FRS [facial recognition systems], you could first of all identify your customers when they entered your store. Then you had the opportunity to identify the emotions they felt as they walked through your store. Of course, some customers keep a "poker face" when they shop, but new software from Californian company, Emotient, can register "micro-expressions"—the tiny flickers of emotion that show on people's faces before they even know they have

registered an emotion or are able to control it. It can even tell you if they are smiling, but not with their eyes.

How useful could this information be when it comes to understanding your customer and creating a unique, personalised experience for them? This kind of granular data is priceless. You can already analyse a customer's online journey through your company website—discovering where they have come from, what they look at, how long they stay on the site, what grabs their attention and what motivates them to action. What if you could apply those same principles to their real-time journey through your real-world store?[44]

Some future-oriented observers of retailing believe that facial recognition will greatly augment existing information-gathering technology at checkout in the not-too-distant future.[45] A Russian technology firm, Synqera, conducted facial recognition trials at checkout in late 2013 for a Russian retail chain. The retailer hoped the activity could supplement loyalty-card information for those customers who have a card, as well as stand in for a loyalty card for those who don't. "If the customer has no loyalty card or doesn't want to identify himself with a loyalty card, then the system recognizes his general mood [by the presence or absence of a smile], gender and age in order to use this data for targeting of the content," said Ekaterina Savchenko, Synqera's head of international marketing. "If the customer identified himself with a loyalty card, the system double checks the customer age and gender with data sourced through facial recognition. If the Synqera system sees that the loyalty card data differs from the camera data, then it evaluates the correctness of the camera data (probability defined for the particular user's gender and age) and, if it is high, gives it priority."[46]

Although theft may be the reason when facial recognition software doesn't match the loyalty card being presented, more likely the card has been loaned to a friend or other family member. In these situations the facial recognition system helps the retailer to parse its data about the name on the card. The checkout systems and the back-end analytics "will then learn that these few users are linked to one card and base the analysis and relevant content on the facial recognition data," Savchenko said.[47] Retail information technology writer Evan Schuman noted that "the biometrics help make sure that the message or promotion being displayed [on a video display at checkout] is the right one."[48] Savchenko said that her firm's software can evaluate whether these are successful for each person at checkout. "Users' smiles are used for the evaluation of the content effectiveness," she said. And users who smile are rewarded, she added: "If the user smiles, he gets a virtual achievement badge or extra loyalty bonuses to his card."[49]

Schuman said that facial recognition systems at checkout can be used either for identifying individual customers by name or, in maintaining anonymity, "merely capturing the facial data points and noting what purchases the person attached to that face makes. Then, when the cameras catch that same face again (say, perhaps four days later), [the system] will remember the prior purchases." With either approach, Schuman said, "such activities need not end with the same channel where they began. Once a shopper is identified in-store and is matched with a CRM [customer relationship management] profile—or they are identified anonymously in-store and a purchase profile of this unknown-person-with-this-specific-face is slowly built—that information can theoretically be married to data from that person's desktop-shopping E-Commerce efforts or their tablet/smartphone's M-Commerce efforts."[50]

Finally, Schuman advanced the possibility that the facial recognition process might be installed outside the physical store as well as inside: "What if [a retail] chain pushes some attractive incentives to get lots of customers and prospects to download its free mobile app? And buried in the terms and conditions is the right for the app to monitor images?" All these recognition approaches would better enable merchants to identify customers entering their store and offer them customized service and deals based on their shopping histories and the retailer's calculation of their long-term value to the business. And while accuracy remains a problem today, Schuman said, "look for this technology to get an order of magnitude more accurate over the next couple of years." As the technology continues to develop, he noted, "the privacy—and associated shopper backlash—risks are obvious." Nevertheless, he contended, "shoppers (especially younger shoppers) seem to have developed an almost infinite capacity for tolerating such efforts. Make the incentive strong enough—and use the data in subtle enough ways so that you're not forcing the customer to know how far you've gone—and privacy will be a trivial concern. Not saying that it *should* be a trivial concern, but merely our belief that it will be."[51] Indeed, over ten years ago retail consultant Karl Bjornson observed that the success of facial recognition systems could well depend on whether the public could be convinced to accept recognition technology as a way to secure their identity and enable them to receive special offers.[52]

Such predictions equally apply to wearable devices. By the mid-2010s retailing consultants were crafting rhetoric to justify data flows from body-hugging technologies. PricewaterhouseCoopers concluded, for example, that "done right, wearable advertising can promise more personal and relevant messaging to consumers—and bring in big business for the companies that leverage this strategically."[53] Similarly, an ad agency CEO proposed continual reinforcement in ways that fit the always-attached nature of the devices: "customizable shopping paths for each person, while incentivizing them with real-time offers and deals along that path."[54] In fact, in an online survey PwC found that millennials more than any other age group would welcome cash, loyalty privileges, and gaming as rewards for adopting wearable devices. The survey found that 37 percent of this age group would be "strongly motivated" to use a wearable device if it "has apps/features that reward those who frequently use it with loyalty points"; 52 percent would be "strongly motivated" if the device "has apps/features that reward those who frequently use it with monetary rewards"; and 64 percent would be "motivated" (though not "strongly motivated") if the device "had some type of gaming component to it."[55]

But just as with the technologies discussed in previous chapters, shoppers who encounter facial recognition systems or who sport wearable devices as they offer up their personal data will likewise not be treated equally. The best customers in the best niches will continue to get the best deals, while, as Bjornson noted, "people not in the right segments will be left behind. They will not have as rewarding an experience."[56] As I've stated throughout, if a primary lesson of the hidden curriculum is that people should get used to giving out their data, another lesson will be that in this great transformation some shoppers will be winners, some will be losers, some will be a bit of both, and many will worry where they stand and why. It is not hard to see that a marketplace centered on individuals' private data and their personalized mobile devices will upend the understanding of shopping that has crystallized over the past century and a half.

Without a doubt, many of the new information technologies created to help people shop online and in physical stores have marvelous capabilities. The ability to carry a device around a store to check shelf prices against those of other stores gives shoppers enormous knowledge—and potential leverage in the purchases they ultimately make. Tailoring a discount coupon to shoppers because of where they happen to be standing in the store at that moment can be a smart way for a retailer to ingratiate itself. Increasingly, though, the data that retailers collect on their own and that they purchase from marketers will be far more personal and will lead to increasingly individualized products, postings, and prices aimed at shoppers. While marketing and retailing executives worry that tailored prices might anger shoppers, they nevertheless agree that discriminatory pricing will become common over time—they consider it an efficient approach to selling in an era of enormous competition. Three of the executives I interviewed argued that offering personalized deals on mobile devices likely would not cause public indignation over undemocratic pricing. These individualized discount coupons, along with tailored product presentations and messages, will be sent just to a shopper's smartphone, tablet, or other privately owned device, so no one else will see them. Shoppers may compare notes about these deals in casual conversation with friends and co-workers, so they may be aware that prices seem to differ for each person, but they will have no clue as to the reasons for the variations, and stores won't provide direct answers.

As I mentioned earlier, the airlines have clearly already paved the way for such discrimination. To begin with, the baroque rules of airline loyalty programs award those considered most loyal with the best seats, first crack at coveted carry-on space, and priority boarding, while others are offered less-desirable seats (though they can shell out more money for better ones), may have to scramble to find a place to stow their carry-on items, and may have to wait longer during the boarding process. There is also the matter of ticket cost, as passengers sitting next to one another on a flight may be paying wildly different fares, and for reasons that are often opaque. There is no record of open rebellion by passengers as a result of discriminatory pricing, seating, storage space, or boarding, and this can be chalked up to the personalized nature of the process: no individual passenger knows what his or her fellow passengers paid for their ticket, or why they received a specific seat or boarding priority. Passengers likely won't discuss these issues among themselves as they sit at the gate waiting to board, partly because of shyness, partly because they don't want to know whether they won or lost in the particular transaction, and partly because they may feel that, whatever the inequalities, those disparities are a legitimate part of the social system. The airlines repeat this message in virtually all communications with their customers: the more you spend and the greater your loyalty, the more you'll get back. In their discriminatory practices they make it clear that fliers who spend less for flights or who fly infrequently aren't as loyal as those who spend and fly more and therefore deserve to be treated less well. You may protest that you simply don't fly frequently for business (a characteristic of the most "loyal" passengers), but to no avail. The airlines have carved out this world of air travel, and we have to live in it.

This type of discrimination-centered world is now evolving in retailing, and it has broader ramifications. In this hypercompetitive environment stores are doing their utmost to steer shoppers to accept that the data-based customer relationship is a natural activity, a process that is taken for granted and a useful part of contemporary life. By doing the right things and having the right purchasing patterns, you will be rewarded with the best product suggestions and the best prices online and in physical stores. Part of "doing the right things" means being a "loyal" customer of the retailers you visit—though being loyal may be more complicated than it sounds. Doing the right thing also means allowing the retailer to gather information about you—for example, by logging in when you visit the merchant's website, downloading (or in some cases just streaming) the retailer's app and allowing location tracking, and enabling Bluetooth so the store can send you coupons as you move through the aisles. It may often mean having the assets and lifestyle that will lead a retailer's computers to conclude that you are a shopper with a high lifetime value to the company.[57]

We are only at the beginning of this retailing transformation. Many of the data collection and tracking technologies that will become standard likely have yet to be invented, and it will take some time before every person experiences these activities on a daily basis. But the trajectory is evident: the technologies of personalization and the social discrimination issues linked to them will become ever central to the ways people shop for goods and services. And the shifting class structure in American society will play a major part, as the changing retail framework will take place against a backdrop of widening income inequality in American society. Beginning in the 1970s income growth for households in the lower and middle classes slowed sharply, while incomes at the top continued to grow strongly.[58] An article sponsored by the University of Southern California's Center on Economic Policy Research concluded that "Americans today live in a starkly unequal society. Inequality is greater now than it has been at any time in the last century, and the gaps in wages, income, and wealth are wider here than they are in any other democratic and developed economy." The article singled out the growing share of recent income gains going to the very high earners (the 1 percent or .01 percent of the population), the emergence of lavishly compensated "supermanagers," and a concentration of wealth that fell a little during the first half of the twentieth century but that has grown steadily since.[59]

One might wonder how many of today's major retailers can survive if such large portions of the population have less real income than in the past. One major way is credit. U.S. middle and lower-middle classes have had to pursue the American dream by borrowing to the hilt. Though the Great Recession of 2008 wrecked many of those ambitions and plunged many people into disastrous debt, there is little evidence borrowing for buying is going away; it simply has become a necessary part of the U.S. economic system. Another way is to target the very wealthy. During the past couple of decades, in fact, retailers and brands have succeeded in the United States when they have pursued both the luxury and the low-price market; the middle is increasingly being hollowed out. There are, in fact, large numbers of people who can afford relatively expensive goods, even if they make up a small percentage of the population overall. This group also includes comparatively wealthy international shoppers, who have increasing access to U.S. merchants through the internet and via stores the retailers have opened abroad. In the process of adapting to these new economic realities merchants have become all the more resolute in their desire to use data and technologies to identify customers of value and pursue them across as many channels as possible. It is increasingly a matter of retailing life and death to fix on and cultivate the right customers while shucking off shoppers who offer little benefit.

The reshaping of retailing will likely add data-driven anxieties to both buyers and sellers. Stores will become stress centers, as technologies aiding shoppers duel with those aiding retailers, resulting in often conflicting conclusions about price and preferential treatment. Sellers will have to change prices constantly, introduce new products rapidly, and continually adopt new ways to define, identify, track, and reevaluate customers, as well as satisfy those they define as winners. Beyond the traditional tension of assessing product quality and cost, shoppers will have to deal with the uncertainty surrounding the personal information that merchants have gathered about them and the scores they are assigned as a result, and the effect that these activities will have on their shopping experience.

This new direction in retail may be healthy for some stores' bottom lines, but it is toxic for people's sense of democratic possibilities in society. The data-driven stratification of customers encourages abandonment of the historical ideal of egalitarian treatment in the American marketplace. As these forms of discrimination accelerate, the retail system increasingly encourages shoppers to accept a coarsening of relations in their dealings with merchants. The old saw that every person's dollar is worth the same increasingly no longer holds sway. Instead, the value of one's dollar depends on the person's prior spending, demographics, lifestyle, and willingness to be an open book for the retailer from whom good treatment and good deals are desired. Certainly the traditional discriminatory categories of race, income, and age will enter these calculations, but proving prejudicial behaviors toward such groups as black, low-income, or older Americans will become far more difficult because they will be masked by complex algorithms. To protect themselves from accusations of bigotry, retailers will probably not even put race into the equation. But it will certainly be represented in a host of hidden factors such as neighborhood, income, education, and even health. And rest assured that shoppers will respond by doing all they can to foster and perpetuate a favorable profile, adjusting their behaviors to those they think will lead to the best treatment from their favorite retailers. Look, too, for whispered social discussions and possibly blogs that try to decipher specific stores' criteria when they pick winners and losers at any given time for any given product.

Jeff Malmad's aggressively utopian declaration that opened this chapter reflects the urgent need that businesses, including retailers, have to gather more and more information on individual consumers. The days when merchants and their advertising agencies targeted broad population segments such as men or college students—or even entire cities—as potential customers are beginning to fade. We are moving rather quickly into an era in which retailers are focused on exploiting what they know about specific individuals to encourage their loyalty and to sell them products. Retailers believe that the more information they secure, the more likely they can identify the hot buttons that computer formulas say will ignite an individual's interest in a particular product. This involves compiling and sifting through huge amounts of data about individuals' shopping habits along with other material (from third-party data firms or from "listening" services) that at first glance doesn't appear directly related to shopping, such as Twitter statements, Facebook comments, blog posts about various products, the names of social media friends, and hobbies. Many merchants agree with IBM that the key to long-term customer profitability is "the process of identifying all of the information for a customer (member data) throughout the enterprise, linking it together for a 360-degree view of a member and maintaining that view going forward."[60]

Of particular concern to many is that much of the data-gathering activities described throughout this book are occurring under the hood. And Malmad's statement that "you'll know everything about yourself and your loved ones if you opt in" inverts what is actually happening: retailers and data providers are trying to learn everything about us and our loved ones whether we know it or not, and if we opt in, an act that is often encouraged by loyalty programs or other incentives, they will learn far more about us than if we don't. Yet Malmad's comments do reflect an intriguing tension that lies at the core of how retailers generally are approaching their customers in the twenty-first century. Even as merchants herald this era as the age of customer power, a result of the hypercompetition brought on by internet trading, they push back against that power via an increasing number of technologies aimed at constraining and channeling the decisions of their customers in ways that benefit the marketers.

This book has charted approaches that the largest department store, supermarket, and big-box chains have adopted to corral desirable customers while minimizing the negative influence of customers who bring them little, if any, profits. Whether these techniques work for the retailers' bottom lines is beside the point; what matters is the larger implications of the emerging retailing institution in which shoppers accept and possibly normalize surveillance and social discrimination. Throughout the book I've noted how emerging technologies and retail marketers' discussions about them turn into taken-for-granted elements of people's worlds. Academic and futurist visions about the ways retailers need to use the new digital technologies to survive on- and offline are translated into practical advice via consultant-generated systems such as customer relationship management, digital intelligence, loyalty systems, and listening platforms. These frameworks, in turn, encourage a continual stream of technologies that put into everyday practice the surveillance, predictive-analytic, and personalization techniques that translate the elite theory into everyday routines. While foes of surveillance and data aggregation in areas such as health, employment, and loan discrimination continue their fight before government agencies, the retailing establishment has built quite similar activities into its very fabric with few complaints. The reason: buying and selling everyday products doesn't seem problematic even when the new tools are used.

And yet, as I've shown, the practice of social discrimination is very much at the core of the transformation of everyday retailing today. The norms of this new retailing era are taking hold: it is rather commonplace that, online and on mobile devices, cookies and their variants form the building blocks of a data-rich world that personalizes offers on the fly. And now online personalization is migrating to the seemingly old-fashioned world of physical retailing—the arena where shoppers still do most of their buying by far. Internet-like tracking and analytical abilities are coming to the aisles and to the checkouts. Tying into the always-on smartphone carried by about 70 percent of Americans, merchants, brand manufacturers, and their agents are exploiting cellular signals, Bluetooth, Wi-Fi, sound waves, light waves, and more to track customers and send them product messages before, during, and after their store visits. The opportunity to receive personally relevant offers is one lure that companies use to entice customers to join the data-collection bandwagon. Enjoying the rewards of protection, privilege, and games relating to loyalty are others.

As a result of the information that retailers gather about their shoppers in this way, the complex algorithms they use mean that they treat people increasingly differently. As they go about their day, including moving through particular stores, people are receiving different messages, even different prices, based on profiles retailers have created about them—profiles that the individuals probably don't know exist and, if they could read them, might not even agree with. While these sorts of discrimination currently take place to a greater degree online than in the physical retailing space, it's clear that the brick-and-mortar establishments are moving inexorably in that direction. Concerns about hypercompetition and notions of efficiency are behind the push, and these changes are overturning more than a century of democratic treatment and posted prices in department stores, groceries, and discount chains. Yet neither the public nor the government has sufficiently grasped the immensity of the changes to be able to ask what they will do to our society.

With these disturbing trends in mind, I hope to encourage a discussion about what can be done to soften, if not eliminate, downsides of this new environment. The data-gathering and data-exploitation activities I've described in this book raise weighty questions. Does the American public really want a society in which marketers are free to track, profile, and target them virtually anywhere they go, and to share information about them with other marketers in ways they do not understand? Are people comfortable with a society that reverts to the discriminatory elements of the peddler era in which selling was based on profiling each customer? If not, how should this new America create room for information respect—the idea that people deserve to know about and manage the ways society's institutions use data about what they say and do?

Marketing, retailing, database, and technology executives typically ignore these questions. They acknowledge the discomfort that some Americans feel regarding the data that firms gather about them.[61] But they claim Americans understand full well they are involved in a trade-off of their data for relevant messages and offers. Marketers depict an informed public that understands the opportunities and costs of giving up its data and makes the positive decision to do so. A 2014 Yahoo! report, for example, concluded that when Americans are online they "demonstrate a willingness to share information, as more consumers begin to recognize the value and self-benefit of allowing advertisers to use their data in the right way."[62] In its argument to policy makers and the media the industry uses this image of a powerful consumer as proof that Americans accept widespread tracking of their backgrounds, behaviors, and lifestyles across devices, even though surveys repeatedly show they object to these activities.

Recall the privacy paradox described earlier in the book, which noted that people have inconsistent and contradictory impulses and opinions when it comes to safeguarding their own private information.[63] Many marketers have embraced this seeming contradiction as a way to argue in favor of wide-ranging data collection. The McCann Worldwide advertising network's Truth Central project derived this conclusion from "a global research study surveying over 10,000 people in eleven countries," including the United States. Neither breaking down the results by country nor detailing the survey method, McCann stated that while "71% worry about the amount online stores know about them, 65% are willing to share their data as long as they understand the benefits for them."[64] The editor for *mCommerceDaily* interpreted the findings to mean that "the tracking of consumers all comes down to the trade-off in value."[65] Along the same lines, the president and chief strategy officer of Mobiquity, a mobile-device-strategy consultancy, wrote in 2012 that "the average person is more than willing to share their information with companies if these organizations see the overall gain for end-users as a goal, not just for themselves."[66] A May 2014 report by Yahoo! Advertising followed this logic in interpreting its survey of "6,000 respondents ages 13–64, a representative sample of the U.S. online population." It highlighted the finding that "roughly two-thirds of consumers find it acceptable or are neutral to marketers using online behavior or information to craft better ads." Digitally connected Americans, the study concluded, "demonstrate a willingness to share information, as more consumers begin to recognize the value and self-benefit of allowing advertisers to use their data in the right way."[67]

Are marketers correct in their assertion that Americans believe in trade-offs and are willing to give up personal data about themselves in favor of discounts and other blandishments? This is a difficult question to answer conclusively. The specific questions asked, and methods used, in these studies often are not included with the survey results, so it's challenging to perform a careful evaluation of the results. Sometimes the respondents are volunteers and their views have no statistical relationship to the population as a whole. A major problem with the accuracy of this research is that the participants are typically recruited online, and it is quite possible that, as opposed to the population at large, volunteers willing to fill out online surveys are comfortable with giving up personal data. And finally, the survey results are inconsistent, and even marketing executives are sometimes loath to fully champion the trade-off view. An Accenture executive interpreted his company's survey to mean that "if retailers approach and market personalization as a value exchange, and are transparent in how the data will be used, consumers will likely be more willing to engage and trade their personal data."[68] The Bain consultancy was even more cautious about its results, saying that "customers' trust cannot be bought by companies offering compensation in exchange for selling or sharing personal data." So what do Americans believe? And if we can determine their views conclusively, what, if anything, should we do about it?

The Annenberg National Internet Surveys, conducted seven times between 1999 and 2015, suggest useful and somewhat provocative answers to these questions. I headed the team at the University of Pennsylvania's Annenberg School for Communication that created each survey. Major polling firms—Roper, ICR, and Princeton Research Associates International—asked the survey questions during twenty-minute interviews with random samples of (typically) fifteen hundred Americans, age eighteen and older. The survey firms contacted prospective respondents by phone—landlines only in the early studies and a combination of cell phones and landlines in recent years to account for the rise of mobile-phone-only households. In our surveys we report the specific questions along with the results (all the questions can be viewed online).[69]

The basic findings are clear and, in some cases, alarming.

1. *Most people know they are being tracked but don't understand what happens behind the screen.* They don't understand data mining, that is, the way companies plumb data about

them from various sources and merge the information to arrive at broad conclusions about them. They also don't understand the purpose of a privacy policy, which is supposed to outline these activities. In four of the surveys (conducted between 2005 and 2015), we asked the participants a slight variation of this question: "True or False: If a website has a privacy policy, it means the site won't share information about you with other sites without your permission." A clear majority of the respondents in all of the surveys believed that this was a true statement, and a substantial percentage beyond that simply said they didn't know. In actuality, the answer is false.

2. *Most people don't know the rules of the new digital marketplace, and they think the government protects them more than it does.* The surveys found that more than half of Americans:

- do not know that a pharmacy does not legally need an individual's permission to sell information to other parties regarding the over-the-counter drugs that the person buys;
- do not know it is legal for an online store to charge different prices to different people buying the identical product at the same time of day;
- do not know it is legal for an offline or physical store to charge different prices to different people buying the identical product at the same time of day;
- do not know that price-comparison sites such as the travel websites Expedia or Orbitz are not legally required to include the lowest travel prices.

3. *Most people don't think personalization by marketers or retailers is a good thing, especially once they know the techniques used for obtaining their personal data.* The 2009 survey conclusively found that, contrary to the claims of many marketers, most adult Americans (approximately 66 percent) do not want to receive tailored advertisements. In contrast, slightly less than half said they would welcome tailored discounts. However, once this group understood several common ways that data are mined to produce tailored discounts—tracking the website the person has just visited, tracking the person on other websites, and tracking the person in stores—much higher percentages said they do not want such messages.

4. *Americans directly admit feeling vulnerable in a retail environment in which companies collect personal information.* The 2005 survey found that more than 70 percent disagreed with the statement "What companies know about me won't hurt me," disagreed with the statement "Privacy policies are easy to understand," and agreed with the statement "I am nervous about websites having information about me." This vulnerable feeling hadn't changed by 2015. We found that 72 percent of Americans rejected the idea that "what companies know about me from my behavior online cannot hurt me."

5. *Most people philosophically do not agree with the idea of trade-offs.* In 2015 we described to a random cross-section of American adults some everyday circumstances whereby marketers collect people's data, and posed this activity in relation to personalized discounts as trade-offs. We found

that far more than half felt those trade-offs were unfair, disagreeing with the statements

- "If companies give me a discount, it is a fair exchange for them to collect information about me without my knowing."
- "It's fair for an online or physical store to monitor what I'm doing online when I'm there, in exchange for letting me use the store's wireless internet, or Wi-Fi, without charge."
- "It's okay if a store where I shop uses information it has about me to create a picture of me that improves the services they provide for me."

But if people disagree with trade-offs on principle, why do they accept them? This apparent contradiction surfaced even in our survey. We presented the people interviewed with a real-life trade-off situation, asking whether they would take discounts in exchange for allowing their supermarket to collect information about their grocery purchases. We found a much higher percentage said yes to the trade-offs than agreed with the three statements above. In examining our findings carefully, we concluded that the reason for this contradiction is resignation, as explained in #6:

6. *Contrary to the claim that a majority of Americans consent to discounts because the commercial benefits are worth the costs, we found that Americans do so because they are resigned to the inevitability of surveillance and the power of marketers to harvest their data.* The meaning of resignation we represent here is, to quote a Google dictionary entry, "the acceptance of something as

undesirable but inevitable."[70] And, in fact, our study revealed that 58 percent of Americans agree with the statement "I want to have control over what marketers can learn about me online," at the same time they agree with the statement "I've come to accept that I have little control over what marketers can learn about me online." Rather than feeling able to make choices, Americans believe it is futile to try to manage what companies can learn about them.

Watching what shoppers do doesn't reveal their attitude. We found that people who believe in trade-offs are quite likely to accept supermarket discounts, but we couldn't predict whether people who are resigned to marketers' data-gathering activities would accept or reject the discounts. Marketers want us to see all those who accept discounts as rational believers in trade-offs. But when we looked at those surveyed who agreed to give up their data for supermarket discounts, we found that well over half of those who took the deal did so out of resignation rather than because they believe in trade-offs.

Ironically, and contrary to many claims for why people give up their personal information, those who are most aware of these marketing practices are more likely to be resigned. Moreover, the more knowledge resigned individuals have of marketing practices, the more likely they are to accept supermarket discounts even when the supermarket collects increasingly personal information. When it comes to protecting personal data, our survey found those with the knowledge to accurately calculate the costs and benefits of maintaining privacy are likely to consider their efforts to do so futile.

Though neither age nor gender reflect any differences in the number of people who are so resigned, statistically significant differences do show up regarding education and race: we found a higher percentage of resignation among the white population compared with nonwhites, and among more educated people compared with respondents who have a high school education or less. At the same time, one-half or more of individuals in those categories of respondents are resigned overall to personal data-gathering by marketers and retailers.

7. *Young adults generally aren't much different than older Americans when it comes to privacy issues.* Media reports teem with stories of young people posting salacious photos online, writing on social networking sites about alcohol-fueled misdeeds, and publicizing other ill-considered escapades that may haunt them in the future. Commentators interpret these anecdotes as representative of a generation-wide shift in attitude toward information privacy. They claim that young people are less concerned than older people when it comes to maintaining privacy. But our research with the Berkeley School of Law found that expressed attitudes toward privacy by American young adults (age eighteen to twenty-four) are not very different from those of older adults. In fact, large percentages of young adults are in harmony with older Americans with respect to both personal privacy and privacy regulations. For example, a large majority of young adults:

- have refused to give information to a business in cases where they felt it was too personal or not necessary;

• believe anyone who uploads a photo of them to the internet should get their permission first, even if taken in public;

• believe a law should be passed to give people the right to know all the information websites have compiled on them; and

• believe a law should be passed requiring websites to delete all stored information about an individual.

In view of these findings, why would so many young adults conduct themselves on social networks and elsewhere online in ways that would seem to forfeit very private information to all comers? One answer is that it's the nature of young adults to approach cost-benefit analyses related to risk differently in comparison with individuals older than twenty-four. An important part of the picture, though, must surely be our finding that higher proportions of eighteen- to twenty-four-year-olds believe incorrectly that the law provides more privacy protection online and offline than it actually does. This lack of knowledge in a tempting environment, rather than a cavalier lack of concern regarding privacy, may be an important reason large numbers of them engage with the digital world seemingly unconcerned.

The Annenberg surveys consistently reflect an American population that is struggling to cope with a media and marketing world they don't understand, one they worry can harm them. Shopping is and will continue to be central to this new world. As U.S. society moves further into the twenty-first century, personalized deals, prices, and other tailored offers will undoubtedly be increasingly troubling to Americans who believe they are on the losing end of often-hidden consumer profiles and targeting formulas. Americans will also surely sense that the changes are causing the democratized marketplace to disappear and will feel powerless to do anything about them.

The issues raised in this book are relevant far beyond the United States. Earlier I discussed a popular Australian marketing consultant who carries his data-personalization boosterism throughout Asia. I also noted Synqera's work on facial recognition in Russian stores. And in 2013 the chief information officer of Tesco, one of the world's largest retailers, linked the key strategic question of how to bring the internet into the stores to three broad forces that would guide his firm's information technology investments: cloud computing, personalization, and the "seamless, blended world of physical and digital." These activities represent the tip of the iceberg of retailing developments involving data collection, tracking, and personalization in-store and via various digital channels that are proceeding at different places around the world.[71] A 2015 Nielsen global report on "The Future of Grocery" begins by describing "a grocery store where you can receive personal recommendations and offers the moment you step in the store, where checkout takes seconds and you can pay for groceries without ever taking out your wallet." Might it sound "far-fetched?" the report asks. Not at all, it answers: "It's closer than you think." The report acknowledges that "today, only a small percentage of consumers around the world is already using such features." But, it adds, this new kind of store is spreading. Moreover, based on a global online survey the company concludes that "willingness to use them in the future is high." It contends that millennials especially are champing at the bit.[72]

The Nielsen report says nothing about surveillance or privacy. Marketing and retailing executives around the world typically have attempted to minimize concerns about their use of shopper data. Our U.S.-focused survey challenges marketers' usual trade-off defense by showing quite clearly that most Americans do not accept the fairness of getting discounts in exchange for their personal data. Yet many in the U.S. marketing industry are making the case that privacy in most areas of shopping doesn't merit concern. They point to controversies about data mining and predictive analytics swirling around the National Security Agency's work as well as around businesses that ferret out targets for shady payday loans and those that dredge up online relationships suggesting a job applicant shouldn't be hired. These intrusions, they insist, are quite different from tracking individuals for the purpose of ordinary advertising and shopping. Consider the comments of Scott Howe, president and CEO of Acxiom:

> Some believe that all data is of equal importance and therefore must be controlled in exactly the same way. This is just not the case. Data regarding personal information that pertains to employment or insurability decisions, or that relates to sensitive health-related issues or confidential matters, deserves much different treatment than data that would indicate that I am a sports fan. Too often, legislation and regulation seek to paint issues in broad strokes, often with unforeseen consequences to liberties and innovation, despite being born of the best intentions.[73]

Apart from ignoring the issue of respecting private information—Americans want to be able to manage their personal data in commerce—this view pushes aside the discriminatory influence that seemingly benign pieces of data can potentially have on an individual's opportunities in the public sphere. Howe surely must have considered that, in the not-too-distant future, companies may well take such actions as merging the category "sports fan" with dozens of other characteristics about individuals—their eating habits (at the ballpark, for example), their income, the number and age of their children, the value of their house, their vacation habits, their geographic whereabouts, their clothes-shopping habits, and their media-use patterns—to create profiles that dub them winners or losers regarding certain areas of shopping and that determine the advertisements and discounts that are directed to them. For reasons they don't understand, people may see patterns of discounts that suggest they are being placed into certain lifestyle segments and are consequently receiving treatment they consider inferior to that of their neighbors and co-workers. The individuals may vaguely understand that these profiles are the cause, and they may try to change their behavior to get better deals, often without success, all the while wondering why "the system"—the opaque predictive analytics regimes that they know are tracking their lives but to which they have no access—is treating them that way.

Arguments that play down the profound discriminatory potential of this sort of data-driven consumer commerce have the same rhetorical aims as those that insist Americans want to exchange data for benefits: they give policy makers false justifications for enabling the collection and use of all kinds of consumer data in ways that the public often finds objectionable. Yet when three of every five Americans are resigned to feelings of powerlessness regarding their data relationships with marketers, when two of every five are both resigned to relinquishing the control of their data to marketers and worried that the loss of control can hurt them, and when people who are well aware of these data-gathering activities are actually more rather than less likely to be resigned, we have a problem. In a central area of the public sphere there exist substantial tensions that cannot be swept away by executives' assertions of consumer autonomy and rational choice.

In 2010 the international retail marketing trade association Point of Purchase Advertising International (POPAI) directly addressed this issue in its "Recommended Code of Conduct for Consumer Tracking Methods." The code concluded that "the ability to record and track a customer's every move through the store, identify customers facially and demographically, and pinpoint where and what customers are looking at, picking up, and putting into their shopping carts through Observed Tracking Data (OTD) raises privacy issues and sends shivers down the spine of even the boldest marketer."[74] By 2015, however, the code had disappeared from POPAI's website, as the marketing and retailing industries had since closed ranks around the idea that these activities are perfectly fine as long as shoppers give their consent by downloading apps and/or turning on Wi-Fi, Bluetooth, location tracking, or other features of their mobile devices that signal a shopper's presence. As we have seen in earlier chapters, loyalty strategies encourage people to opt into tracking via apps. We have also seen that retailers create privacy policies that intentionally avoid shedding any light on tracking activities. The policies not only are turgid documents designed to be unreadable by nonexperts; they are tough-luck contracts draped in take-it-or-leave-it terms. Moreover, their opaque nature decreases the likelihood that shoppers will become upset over specific details of the merchants' use of their data. Between the policies' true intent and our finding that most Americans assume a "privacy policy" means that their data won't be shared without their permission—and the fact that the FTC has made no attempt to implement any changes that would clear up this misconception—we have a brick-and-mortar retailing environment that is a frontier for customer data exploitation.

The marketing and retailing establishments generally have been secure pursuing their activities without government intrusion. A notable exception is the Children's Privacy Protection Act, a landmark law that makes it illegal for companies to gather data on, or track, children under the age of thirteen without explicit parental permission. This one law notwithstanding, Tony Hadley, CEO of Experian, which sells both online and offline information on hundreds of millions of Americans, stated confidently in 2014 that Congress would not pass a law regulating the collection of marketing data in the foreseeable future, or maybe ever. "Should marketing data be regulated like credit, like employment, like lending data?" he pondered, referring to specific laws that address financial and health data in the name of privacy. "I think the clear indication that we're getting from Congress is no."[75]

Beyond Congress, the Federal Trade Commission (FTC) has taken the lead at the national level in wading into the controversies surrounding the activities that Experian and so many of the retailers and intermediaries are pursuing. A number of the agency's commissioners and staff members are sympathetic to complaints expressed by individuals and advocacy organizations regarding potential misuse of digital-marketing data. In an attempt to prevent deception, the FTC in 1998 adopted a set of "fair information practice principles" that require firms to provide "notice" of their information practices; "choice" (giving individuals options about how their personal identifying information is used beyond the initial purpose); "access" (giving individuals reasonable opportunity to review the information collected and to make corrections); and "security" (reasonable steps to protect the information). Over time, the FTC has added to these principles with categories such as "collection limitation," "purpose specification," and "privacy by design." Yet privacy expert Robert Gellman has concluded that, "from 1998 through 2010, the Commission's description of [fair information practice principles] has been consistently inconsistent."[76] More troubling, companies have learned how to conform to the letter of the FTC categories while running roughshod over their spirit—the incoherent privacy policies, a response to the FTC requirement that websites provide "notice" of their information practices, are a primary example. Marketers and retailers also play games with the "choice" and "access" principles; for example, the typical tough-luck privacy policy is an illusion of choice.

Marketers have generally had free rein to track people and collect information about them, as the FTC has been hobbled by the narrow definition of "harm" that it (and advocates petitioning it) must follow before it can act: injury to a person resulting from the use of the person's gender, race, or age in health care, financial, or employment decisions and ending in monetary loss. For example, the FTC has shown an interest in whether certain loan companies collect personal data that, while seeming to have nothing to do with race, enables those companies to circumvent race discrimination laws. The commission has also been willing to penalize companies that have gathered or have used data in a deceptive manner. Although these are important initiatives, the FTC's definition of harm prevents it from addressing fundamental concerns about the public's right to know about the specific ways companies compile personal data and how they use the data. Nor can it mandate limits on in-the-aisle data gathering in the new retailing era. As the FTC itself noted in a 2010 report, approaches to protecting consumer privacy that employ a harm-based model tend to focus on enhancing physical security, limiting economic injury, and minimizing unwanted intrusions into the daily lives of consumers.[77] The report goes on to note that such an approach tends to leave out reputational harm and the pervasive fear many Americans have of being watched—issues that certainly intersect with concerns we have found in the Annenberg surveys. Yet the issues our surveys have identified are both more specific and at the same time broader: Americans want to have control over the data that marketers obtain about them. They believe that marketers currently have total control, and they worry (perhaps because of that lack of autonomy) that the commercial relationships they have as a result of marketers possessing their personal information are fraught. And they feel resigned to being unable to do anything about it.

The retail transformation taking place challenges the usefulness of the fair information practices and other self-regulatory regimes set up to protect America's right to privacy in the commercial realm. Philosopher and New York University professor Helen Nissenbaum notes that privacy involves the contextually appropriate flow of personal information,[78] meaning that this flow should be consistent with the expectations of those whose data is being used. A website's "notice" of its information practices helps individuals to determine whether the flow meets their expectations. Yet the evolving cross-platform surveillance activities of the new retailing world make it difficult, if not impossible, for an individual to understand and act on even this one aspect of the fair information practice principles; the person would continually have to review and attempt to comprehend several sets of frequently changing privacy practices of merchants online, on apps, and in the physical stores. They would also have to be familiar with the evolving privacy practices of third-party app companies such as shopkick and inMarket. Even in a perfect world, carrying out such due diligence successfully would be a huge task.

Companies such as Google and Acxiom say they offer individuals the opportunity to view and correct or update their personal information stored in the businesses' database. However, these firms join with other companies to meld their respective sets of data in targeting potential customers, resulting in individualized messages that often incorporate far more personal information than the characteristics individuals see. Further, in 2013 the *New York Times* reported that the website Acxiom had created to make this information available, aboutthedata.com, wasn't releasing all the personal information it had collected.[79] Jeff Chester of the consumer group Center for Digital Democracy told the newspaper that the website's language "is so innocuous that the average consumer would think there's no privacy concern."[80] A comparison between the information Acxiom was making available to individuals on aboutthedata.com and the descriptions of what it was selling in its 2013 *Consumer Data Products Catalog* indicates that as of late 2015 the company was still making available only a small percentage of the personal data. For example, the website did not allude to the company's tracking of people on social media, yet the catalog offers a panoply of products that provide such information as the number of an individual's social network friends and the names of the people an individual is following—and is followed by—on Twitter.[81] Health data is another example, as the website did not mention that Acxiom collects this type of personal data yet the product catalog sells information focusing on such areas as an individual's cholesterol level, diabetes, or senior needs. Nor did the website make any mention of the sixty-seven profiling tags the catalog assigned to people based on household socioeconomic status, such as "Summit Estates," "Humble Homes," and "Resilient Renters."

Acxiom's actions suggest a misleading approach that would keep even the most assiduous people from understanding the activities surrounding their personal information, as do those of Google and other firms. Although the Federal Trade Commission is mandated to monitor and punish deception in the marketplace, one has to wonder whether the fair information practice principles themselves actually aid deception or misleading inferences by pretending that notice, choice, and access can actually be carried out in the emerging complex, multilayered world.

* * *

What is taking place is clearly not business as usual. It's not even the business that existed when the fair information practice principles were written. We cannot rely on self-regulation by marketers and the monitoring abilities of individual shoppers to resolve the complex ethical, social, and legal problems created by the new regimes of commercial surveillance. Our Annenberg surveys consistently indicate that even when people consent for whatever reason to being tracked, or to allow retailers to use their personal information, they don't truly understand what's taking place behind the screen. At the heart of the situation is an asymmetry of power: on the one hand, retailers have vast amounts of personal information on people and can identify them using a range of means, and on the other, the public has little recourse in educating itself and acting on those activities.

Regulation is a crucial means for addressing this situation, as it can immediately slow the pace of marketing surveillance. Some public-interest advocates frustrated with the FTC hold out hope that the Federal Communications Commission (FCC) will step up with stricter regulations that encourage transparency of data use by marketers and discourage tracking in the commercial sphere without permission. Unlike the FTC's mandate to counter business harms, the FCC has the broader government mandate to regulate "in the public interest, convenience and necessity," to quote the 1936 Federal Communications Act. In recent years, the FCC has reclaimed the right to regulate internet service providers (ISPs) as common carriers, similar to phone services. In 2016 FCC chairman Tom Wheeler suggested an FCC edict that those providers (typically cable firms and large telecommunications firms such as AT&T and Verizon) should be prohibited from sharing customer data without active ("opt-in") permission by those individuals. Wheeler reasoned that because they monitor all the broadband traffic to every location and device, ISPs have the capability of learning enormous amounts about where their customers go and what they look at online and on apps. Public interest activists cheered the idea, but the ISPs reacted angrily. They argued the FCC shouldn't be regulating them with draconian rules while, they contended, the FTC was allowing Google, Facebook, and other commercial giants to scarf up at least as much data about individuals. The ISPs gathered the support of important lawmakers, who announced the FCC was overreaching its regulatory authority. As of late 2016 Wheeler remained intent on pushing his plan through the FCC, and his opponents remained equally intent on making sure such opt-in rules never get implemented.

Rather than pursuing the ISP argument against opting in, the correct approach is actually to argue for the opposite: require a comprehensive opt-in policy for every company that wants to use an individual's data, whether it be Google, inMarket, or Macy's. Retailers will object, stating that people already opt in to being tracked when they install apps on their mobile devices, but this is a disingenuous argument requiring one to seriously believe that individuals stop to read the legalese of privacy policies on their smartphones' small screens at the moment of download. One way such an opt-in could be implemented for apps is to prohibit the company offering the app from tracking individuals immediately after download. Instead, the firms should send downloaders a straightforward accounting of data use through email (under the condition that the firms discard the email addresses afterward). Only after the downloader responds affirmatively to the email, or uses the app at least a day after receiving the email, should the app owner be allowed to husband data.

These opt-in requirements clearly will not address the many issues regarding commercial surveillance in the imminent omnichannel environment. But they can slow the growth of surveillance in the aisles so that society can then have the necessary time to chart a retailing future that satisfies marketers but protects people from being overrun by surveillance and minimizes discrimination. Despite the insistence of retailers' lobbyists, we can't put the burden on shoppers, who have a life and don't have time to learn the ins and outs of new technologies that are often sugarcoated by companies that have a vested interest in deceiving them. The term *transparency* is used a lot in the marketing trade press to signify advertisers' insistence that they must be able to look into the specifics of their programmatic-advertising activities so that they can calculate their return on investment and not get harmed monetarily. Yet when it comes to transparency in advertisers' relationships with the public, the term is far less popular—and the activities related to implementing openness far less diligent. We need initiatives that give the public the right and ability to learn what companies know about them and how they profile them, and what forms of data lead to what kinds of personalized offers. We also need to get people excited about using that right and ability. Here are a few suggestions:

- *Encourage more corporate openness about the commercial use of people's data by naming, praising, and shaming.* Public interest organizations as well as government agencies should develop clear definitions of transparency that reflect concerns identified in the Annenberg surveys and elsewhere. They should then systematically identify companies accordingly. When activists, journalists, and government officials name and shame firms that don't abide by the transparency norms, they can alert the public to stay away from bad actors, possibly force those actors to change their behaviors, and encourage some firms to see privacy as a selling point.

• *Create an initiative that dissects and reports on the implications of privacy policies.* Activists, journalists, and government officials—perhaps aided by crowd-sourcing initiatives—should take on the role of interpreting these legally binding documents for the public. Rather than focusing on whether websites abide by their privacy policies, privacy policy interpreters can be most helpful by uncovering how companies say they collect and use their data, and what the implications might be for the individual and for society. They should also advocate for a consumer right to agree with selective parts of privacy policies by underscoring policies that do allow this. When this information is available in a digestible form, it may spur informed naming, praising, and shaming. It may well also lead firms to alter objectionable behaviors.

• *Give individuals the right to know the specific data and profile on which a retailer bases any of its targeted messages, coupons, or other interactions.* As long as the algorithms companies implement to analyze and predict the future behaviors of shoppers are hidden from public view, the potential for unwanted marketer exploitation of individuals' data remains high. We therefore ought to consider it an individual's right to access the profiles and scores companies use to create every personalized message and discount that the person receives. Although companies will argue that giving out this information exposes trade secrets, we contend that this can be done without exposing companies to such damage.

• *Educate the public about digital media and marketing, beginning in middle school.* To have a truly informed discussion, people need to be able to communicate in the language of the future. They have to learn the vocabulary of digital media and marketing and be familiar with the primary individuals behind it. This area should be considered a part of the liberal arts because a solid understanding of it is necessary for a thriving citizenry.

Activists, journalists, and schoolteachers all need to be educators when it comes to the changing retailing institution and its hidden curriculum. They can push against the asymmetry of power in retailing by casting light on the often non-privacy-oriented nature of merchants' privacy policies, focusing on how companies say they collect and use their data, and on what the implications might be for the individual and society. When these practices are available in digestible form—and people realize how different the retailers' norms are from their own—the knowledge may incite the praising or shaming of retailers and result in chastened merchants that change the behaviors their shoppers find disagreeable. These public activities, in turn, may spur the FTC, the FCC, Congress, state regulators, and the courts to work toward as even and as open a shopping playing field as possible. Also helping to push in this direction, and to puzzle out some of the legal conandrums the new environment is creating, are the platoons of privacy and surveillance scholars the nation is fortunate to have.[82] They offer important ideas about regulation, corporate behavior, and cultural values that might well help regulators confront the evolving retail challenges highlighted in this book.

The stakes are high. Our society should not rush headlong into a new retailing world, but instead should question whether it is the one we want. Should payment for products include a part of yourself? Do you want the next generation strolling down store aisles and thinking it normal that the merchants have profiled and scored them, often in prejudicial ways, and that they don't know how they were labeled, what consequences the labeling will have on their shopping experience, and whether they have any say in the matter? Merchants, left to their own interests and in response to hypercompetition, are creating this world. And they continue to work with the digital industry, which has grown around them, to ensure that future generations accept an environment of surveillance and tracking. If the retail industry prevails, a future societal mantra might well be, "Shoppers want to be tracked." Our descendants might well remember when we failed to make a choice—if they recall there was a choice at all.

ACKNOWLEDGMENTS

This book explores a new retailing industry in the making, and its effect on how we see ourselves and our world. I owe a lot to the many marketing executives, agency practitioners, journalist-experts, and public interest advocates who answered my questions about the complex technologies, business processes, and government policies that are guiding the transformation. Several of these generous individuals wished to remain anonymous. I am thankful for the opportunity to acknowledge the others together with the firms where they were employed when we talked: Bruce Biegel (Winterberry), Bill Bishop (Willard Bishop), Michael Boland (BIA/Kelsey), Ryan Bonificina (Alex and Ani), Jeff Chester (Center for Digital Democracy), Andy Chu (Sears), Ryan Craver (Hudson's Bay), Abhi Dhar (Walgreens), Pam Dixon (World Privacy Forum), Ivan Frank (Taubman), Jason Goldberg (Razorfish), Ethan Goodman (The Mars Agency), Moshe Greenshpan (Face-Six), Jeff Griffin (inMarket), Zach Grossman (Perka), Michael Healander (GiSi), Laura Heller (Fierce Retail), Kevin Hunter (industry consultant), Jeff Jensen (inMarket), Randy Jiusto (Outsell), Matthew Kulig (Aisle411), Daniel Mahl (Thinknear), Bill Martin (ShopperTrak), Charles Martin (Umbel), Michael Miller (Catapult), Patrick

Moorhead (Twitter/Catalina), Brett Reisman (Spreo), Suzi Robinson (Stop & Shop), Joe Stanhope (Signal Digital), Mark Tack (Vibes), Paul Verano (Shopperception), Tyler Watson (inMarket).

In many cases I contacted sources because they either were mentioned in or wrote trade magazine articles I found helpful, or because they appeared on panels at industry meetings I attended. Particularly useful sources for tracking contemporary developments were MediaPost's *Email Marketing Daily, Mobile Marketing Daily, Search Marketing Daily, Social Media and Marketing Daily, Real Time Daily,* and *Media Daily News;* Fiercemarket's *Fierceretail, FierceMobileMarketer, FierceMobileRetail,* and *FierceBigdata; Mobile Commerce Daily, Progressive Grocer, Advertising Age, AdWeek,* and *Adexchanger.* The industry meetings were useful for confirming, extending, or refuting what I had learned in the trades. Steve Smith, Joe Mandese, and Wendy Davis at MediaPost Communications were particularly generous in allowing, even inviting, me to attend a number of their conferences and making me feel welcome. I benefited from reports from retailing and marketing consulting and research firms such as Forrester Research, eMarketer, Winterberry, Accenture, IBM, and Jainrain.

Colleagues from various academic disciplines have been crucial to teaching me, critiquing me, and offering me sage advice about various aspects of marketing, retailing, the digital environment, surveillance, and privacy. Warm thanks are due Danielle Citron, Brooke Duffy, Chris Jay Hoofnagle, Kathy Montgomery, Monroe Price, Ira Rubinstein, Christopher Yoo, and the always enlightening speakers and papers at the annual Privacy Law Scholars conference staged on alternate years by Chris Hoofnagle and Dan Solove at Berkeley Law School and George Washington

University Law School. My thinking on topics in this book has also benefited from research help by Annenberg doctoral students Ope Akanbi, Nora Draper, Emily Hund, Elena Maris, Lee McGuigan, Paul Popiel, John Remensperger, and Steven Schrag.

I am fortunate to work in an environment that provides the time and encouragement to carry out this sort of research. Michael Delli Carpini, dean of University of Pennsylvania's Annenberg School, has had a large role in creating and perpetuating that atmosphere, and I thank him for it. Thanks also go to Joe Calamia, my editor at Yale University Press. Joe's enthusiasm for this project and sensitive editorial suggestions have made the process of getting this book out as smooth as it could possibly be. The work also benefited from the talents of Jeff Schier. A master copyeditor, Jeff worked hard to bring out the best in every paragraph.

NOTES

1 A FROG SLOWLY BOILED

1. Conservative estimate of Kevin Hunter, an industry consultant who until July 2016 was president of the inMarket beacon network.
2. See Brandon Fischer, "Bright and Shiny Objects", http://www.slideshare.net/mediapostlive/brandon-fischer, accessed September 25, 2015.
3. http://www.fmediapostlive/brandon-fischer, published August 7, 2015, slide 12i.org/research-resources/supermarket-facts; http://www.fmi.org/research-resources/supermarket-facts/weekly-household-grocery-expenses-2014.
4. http://www.nacsonline.com/Research/FactSheets/ScopeofIndustry/Pages/Convenience.aspx; Cushman and Wakefield Research, "Reality Check: Evaluating 7 Industrial Real Estate Predictions," October 2014, http://global.cushmanwakefield.com/en/research-and-insight/2014/reality-check-industrial-re-predictions-2014/ (page 4).
5. Shelley Banjo and Drew Fitzgerald, "Stores iConfront New World of Reduced Shopper Traffic," *Wall Street Journal*, January 16, 2014, http://online.wsj.com/news/articles/SB10001424052702304419104579325100372435802, accessed August 22, 2014; Cushman and Wakefield Research, "Reality Check: Evaluating 7 Industrial Real Estate Predictions," October 2014, http://global.cushmanwakefield.com/en/research-and-insight/2014/reality-check-industrial-re-predictions-2014/ (page 4).
6. Ibid.
7. Marco Kesteloo and Nick Hodson, "2015 Retail Trends," January 13, 2015, http://www.strategyand.pwc.com/reports/2015-retail-trends;

"On Solid Ground: Brick-and-Mortar Is the Foundation of Omnichannel Retailing," ATKearney, 2014, https://www.atkearney.com/documents/10192/4683364/On+Solid+Ground.pdf/f96d82ce-e40c-450d-97bb-884b017f4cd7, accessed February 21, 2016.

8. Elizabeth A. Harris, "GE Capital to Help Set Up Loyalty Programs for Retailers," *New York Times,* January 13, 2014, http://www.nytimes.com/2014/01/13/business/ge-capital-to-help-set-up-loyalty-programs-for-retailers.html?_r=0, accessed March 17, 2014.
9. Laura Heller, "Consumers Love Loyalty; Will They Stay Loyal to Starbucks?," Fierce Retail, March 2, 2016, http://www.fierceretail.com/story/consumers-love-loyalty-will-they-stay-loyal-to-starbucks/2016-03-02, accessed March 2, 2016.
10. Charles Duhigg, "How Companies Learn Your Secrets," *New York Times Magazine,* February 16, 2012, http://www.nytimes.com/2012/02/19/magazine/shopping-habits.html?pagewanted=all&action=click&module=Search®ion=searchResults%230&version=&url=http%3A%2F%2Fquery.nytimes.com%2Fsearch%2Fsitesearch%2F%3Faction%3Dclick%26contentCollection%3DBusiness%2520Day%26region%3DTopBar%26module%3DSearchSubmit%26pgtype%3Darticle%23%2Ftarget+pregnant, accessed March 17, 2014.
11. Jane Martin, "What Should We Do with a Hidden Curriculum When We Find One?" *Curriculum Inquiry* 6:2 (1976): 135–51.
12. Emile Durkheim, quoted in Eric Margolis et al., "Hiding and Outing the Curriculum," in *The Hidden Curriculum in Higher Education,* ed. Eric Margolis (New York and London: Routledge, 2001), 1–20.
13. Roland Meighan, *A Sociology of Educating* (New York: Holt, Rinehart, and Winston, 1981), 52.
14. Jean Anyon, "Social Class and the Hidden Curriculum of Work," *Journal of Education* 162:1 (Winter, 1980): 67–92.
15. Samuel Bowles and Herbert Gintis, *Schooling in Capitalist America* (New York: Basic Books, 1976), 42; see also 163.
16. Paul Willis, *Learning to Labor* (New York: Columbia University Press, 1997).
17. Thomas Berger and Peter Luckman, *The Social Construction of Reality* (New York: Anchor Books, 1966), 55.
18. Ibid., 42, 62–63; see also 24.

19. Interview with retail executive, September 21, 2014.
20. The term *social imaginary* is borrowed from social philosopher Charles Taylor. See Charles Taylor, *The Secular Age* (Cambridge, MA: Harvard University Press, 2007), 1.
21. Brian Kilcourse, RSR research, quoted in Leonie Barrie, "NRF 2014: Tailoring Technology for All the Omnichannel Success," just-style, January 21, 2014, http://www.just-style.com/analysis/tailoring-technology-for-omnichannel-success_id120413.aspx, via LexisNexis.
22. Berger and Luckman, *The Social Construction of Reality*, 55.

2 THE DISCRIMINATING MERCHANT

1. "Trust Legislation: Hearings Before the Committee on the Judiciary House of Representatives, 63rd Congress, On Trust Legislation, Serial Seven—Parts 1 To 10 Inclusive, Volume 2," Washington: Government Printing Office 1914, 731.
2. Claire Holleran, *Shopping in Ancient Rome* (Oxford, UK: Oxford University Press, 2012), 240.
3. Laurence Fontaine, *History of Pedlars in Europe* (Durham, NC: Duke University Press, 1996), 91.
4. Ibid.
5. Tracey Deutsch, *Building a Housewife's Paradise: Gender, Politics, and American Grocery Stores in the 20th Century* (Chapel Hill, NC: University of North Carolina Press, 2010), 33.
6. Ibid.
7. Ibid., 33–40.
8. Robert Hendrickson, *The Great Emporiums* (New York: Stein and Day, 1979), 28.
9. Ibid., 29.
10. Ibid., 31.
11. William Leach, *Land of Desire: Merchants, Power, and the Rise of a New American Culture* (New York: Vintage, 1993), 6.
12. Hendrickson, *Great Emporiums*, 31.
13. Mica Nava, "Modernity's Disavowal: Women, the City, and the Department Store," in *The Shopping Experience*, ed. Pasi Falk and Colin Campbell (Thousand Oaks, CA: Sage, 1997), 60.

14. Ibid.
15. Barbara Olsen, "Rethinking Marketing's Evolutionary Paradigm," in *Explorations in Consumer Culture Theory,* ed. John F. Sherry and Eileen Fischer (New York: Routledge, 2008), 65.
16. Nava, "Modernity's Disavowal," 64.
17. Ibid., 66.
18. Emile Zola, *The Ladies' Delight,* trans. Robin Buss (London: Penguin Classics, 2006), 75.
19. Nava, "Modernity's Disavowal," 66.
20. Harold W. Fox, *The Economics of Trading Stamps* (Washington DC: Public Affairs Press, 1968), 32.
21. Hendrickson, *Great Emporiums,* 36.
22. Quoted in Leach, *Land of Desire,* 112.
23. Leach, *Land of Desire,* 133.
24. Ibid., 133.
25. Susan Strasser, *Satisfaction Guaranteed: The Making of the American Mass Market* (New York: Pantheon, 1989), 248.
26. Deutsch, *Building a Housewife's Paradise,* 57.
27. Daniel Delis Hill, *Advertising to the American Woman, 1900–1999* (Columbus, OH: Ohio State University Press, 2002), 3–5.
28. From the *Illustrated Weekly* caption, as reproduced in Montrose Morris, "Walkabout: The Great Milk Wars, Part 1," The Brownstoner, November 8, 2011, http://www.brownstoner.com/history/walkabout-the-great-milk-wars-part-1/, accessed September 25, 2015.
29. Fox, *Economics of Trading Stamps,* 34.
30. Ibid.
31. Deutsch, *Building a Housewife's Paradise,* 144.
32. Fox, *Economics of Trading Stamps,* 34.
33. Deutsch, *Building a Housewife's Paradise,* 144–45.
34. Ibid., 145.
35. Quoted in Deutsch, *Building a Housewife's Paradise,* 146.
36. Ibid.
37. Deutsch, *Building a Housewife's Paradise,* 147–48.
38. Ibid., 187.
39. John Conner, "Supermarkets: There'll Be Another One Along Any Minute," *Collier's,* May 1951, 64.

40. Tom Mahoney and Leonard Sloan, *The Great Merchants: America's Foremost Retail Institutions and the People Who Made Them Great* (New York: Harper and Row, 1974), 17.

41. Ibid.

42. Leach, *Land of Desire*, 6.

43. Deutsch, *Building a Housewife's Paradise*, 192.

44. "Supermarkets as Symbols," National Museum of American History, http://americanhistory.si.edu/food/new-and-improved/supermarkets-symbols, accessed September 21, 2015.

45. Susan Porter Benson, *Counter Cultures: Saleswomen, Managers, and Customers in American Department Stores, 1890–1940* (Urbana, IL: University of Illinois Press, 1986), 91.

46. Ibid., 89–90.

47. Ibid., 90.

48. Ibid., 102.

49. Leach, *Land of Desire*, 133.

50. Deutsch, *Building a Housewife's Paradise*, 68.

51. Ibid., 213.

52. "Supermarkets' Newest Special: Self Defense," *Business Week*, September 7, 1968, 34.

53. Deutsch, *Building a Housewife's Paradise*, 216.

54. Benson, *Counter Cultures*, 114.

55. Nielsen Company, "Celebrating 90 Years of Innovation," http://sites.nielsen.com/90years/.

56. Sarah E. Igo, *The Averaged American: Surveys, Citizens, and the Making of a Mass Public* (Cambridge, MA: Harvard University Press 2007), 212.

57. Benson, *Counter Cultures*, 114.

58. Ibid., 113.

59. Ibid.

60. Interv ew with William Bishop, September 2014.

61. Paul . Nystrom, *Retail Selling and Store Management* (New York and London: D. Appleton, 1916), 35.

62. Edward A. Filene, *The Model Stock Plan* (New York: McGraw-Hill, 1930).

63. Deutsch, *Building a Housewife's Paradise*, 67.

64. Paul H. Young, "Opportunities for the Use of Research in the Management of a Retail Department Store" (Thesis in Marketing: The Wharton School of the University Of Pennsylvania, 1946), 46.

65. Faye Gold, Alfred E. Berkowitz, Milton M. Kushins, and Edward A. Brand, *Modern Supermarket Operations,* 3d ed. (New York: Fairchild, 1981), 112.

66. Ibid.

67. Fox, *Economics of Trading Stamps,* 1.

68. Ibid.

69. "Ask for S. & H. Green Trading Stamps at Davidson's Cash Store," advertisement in *Arizona Republican,* April 26, 1910, 7.

70. Fox, *Economics of Trading Stamps,* 94.

71. Jeff R. Lonto, "The Trading Stamp Story," Studio Z.7 Publishing, http://www.studioz7.com/stamps.html, accessed November 1, 2014.

72. Fox, *Economics of Trading Stamps,* 36.

73. George Church, "Give-Away Selling Tool Draw New Grocer Ire," *Wall Street Journal,* April 13, 1956, 8.

74. Lonto, "Trading Stamp Story."

75. Warren Buffett, "To the Shareholders of Berkshire Hathaway, Inc.," Berkshire Hathaway, February 28, 2007, http://www.berkshire hathaway.com/letters/2006ltr.pdf, accessed September 18, 2015.

76. Lonto, "Trading Stamp Story."

77. John T. Dunlop and Jan W. Rivkin, Introduction to Stephen A. Brown, *Revolution at the Checkout Counter* (Cambridge, MA: Harvard University Press, 1997), 23.

78. Ibid.

79. Ibid., 24.

80. Lonto, "Trading Stamp Story."

81. Randy Alfred, "June 1974: By Gum! There Is a New Way to Buy Gum," *Wired,* June 26, 2008, http://archive.wired.com/science/discoveries/news/2008/06/dayintech_0626?currentPage=all, accessed on November 1, 2014.

82. Dunlop and Rivkin, Introduction, 22–23.

83. Alfred, "June 1974."

84. Dunlop and Rivkin, Introduction, 6.

85. Ibid., 5.

3 TOWARD THE DATA-POWERED AISLE

1. Robb Mandelbaum, "Small Businesses and Amazon," *New York Times,* December 15, 2011, B5.

2. Richard Russo, "Amazon's Jungle Logic," *New York Times,* December 13, 2011, A35.

3. "Amazon.com's Stunt," *Lowell Sun,* December 9, 2011, editorial page.

4. Stephanie Clifford and Claire Cain Miller, "Rooting for the Little Guy," *New York Times,* January 6, 2012, B1.

5. Bruce Dadey, quoted in John Barber, "Amazon's Showrooming Tactic Enrages Booksellers," *Globe and Mail,* December 24, 2011, R12.

6. Mandelbaum, "Small Businesses and Amazon."

7. Stephanie Clifford and Julie Bosman, "Target, Unhappy with Being an Amazon Showroom, Will Stop Selling Kindles," *New York Times,* May 3, 2012, B3.

8. Marilyn Much, "Brick-and-Mortar Chains Like Macy's Battle vs. Amazon," *Investor's Business Daily,* May 15, 2012, A1.

9. Clifford and Bosman, "Target, Unhappy with Being an Amazon Showroom, Will Stop Selling Kindles."

10. Isadore Barmash, "Marketplace; Investors Ignore Wal-Mart Gains," *New York Times,* June 20, 1988, D6.

11. Ibid.

12. Eben Shapiro, "3 Discounters on a Collision Course," *New York Times,* September 23, 1991, D1.

13. Thomas C. Hayes, "Behind Wal-mart's Surge, a Web of Suppliers," *New York Times,* July 1, 1991, D1; Leslie Kaufman, "Wal-Mart: Retail Showcase Some Hate," *Contra Costa Times,* October 29, 2000, D1.

14. Associated Press, "The 400 Richest People," *Los Angeles Times,* October 13, 1987, http://articles.latimes.com/1987-10-13/business/fi-13869_1_real-estate, accessed December 12, 2014; *New York Times,* October 14, 1986, D6.

15. Marybeth Nibley, "Wal-Mart Plans Business Smarts, Small-Town Charms," *Palm Beach Post,* March 26, 1989.

16. Ibid.

17. Ibid.

18. Christopher Sullivan, "Wal-Mart Is High-Tech on Low Retail," *Hamilton Spectator* (Ontario, Canada), September 23, 1993, F7.

19. Ellis Booker, "IS Trailblazing Puts Retailer on Top," *Computerworld*, February 12, 1990, 69; Amy Helen Johnson, "A New Supply Chain Forged," *Computerworld*, September 30, 2002, 38.

20. Booker, "IS Trailblazing Puts Retailer on Top"; Johnson, "A New Supply Chain Forged."

21. Tony Seidman, "Toy Industry Takes Big Steps in Automation," *Journal of Commerce*, February 13, 1989, 4B.

22. Booker, "IS Trailblazing Puts Retailer on Top."

23. Charles Russnell, "They Know How Shoppers Think," *Ottawa Citizen*, April 26, 1994, D3.

24. Ibid.

25. Robin Bulman, "Logistics System Propelled Wal-Mart to Leading Role in Retail Industry," *Journal of Commerce*, April 19, 1994, 11C.

26. Tim Shorrock, "Retailers Push Down Prices, Alter Sourcing in Apparel Sector, Forum Told," *Journal of Commerce*, May 21, 1997, 3A.

27. Kaufman, "Wal-Mart: Retail Showcase Some Hate."

28. Ibid.

29. Ibid.

30. Ibid.

31. Ibid.

32. Ibid.

33. Russnell, "They Know How Shoppers Think."

34. Bulman, "Logistics System Propelled Wal-Mart to Leading Role in Retail Industry."

35. Lorrie Grant, "An Unstoppable Marketing Force Wal-Mart Aims for Domination of the Retail Industry—Worldwide" *USA Today*, November 6, 1998, 1B.

36. Julia King, "Can America Win the Wardrobe Wars?" *Computerworld*, January 24, 1994, 67; Kaufman, "Wal-Mart: Retail Showcase Some Hate."

37. Leslie Wayne, "Rewriting the Rules of Retailing," *New York Times,* October 15, 1989, 3/1.
38. "Industry Snapshot," *Investor's Business Daily,* September 14, 1998, A33.
39. Ibid.
40. Ellen Neuborne, "A Retail Giant in Store: Federated to Acquire Rival Macy," *USA Today,* July 15, 1994, IB.
41. Jenny McTaggart, "The New World Order," *Progressive Grocer,* August 2012.
42. Jennifer Lawrence, "Cola Wars Move In-Store," *Advertising Age,* November 9, 1987.
43. Julie Liesse, "EDLP Leaves Trade in Big Flux; Harsh New Reality for Retailers," *Advertising Age,* May 10, 1993, 4.
44. Judann Dagnoli, "P&G Plays Pied Piper on Pricing," *Advertising Age,* March 9, 1992.
45. Julie Liesse, "EDLP Leaves Trade in Big Flux; Harsh New Reality for Retailers."
46. McTaggart, "The New World Order."
47. Alan J. Ryan, "Retailers Seek Systems Edge," *Computerworld,* October 23, 1989, 23.
48. Lucie Juneau, "Luring Consumers with Conspicuous Efficiency," *Computerworld,* September 14, 1992, 37.
49. Neuborne, "A Retail Giant in Store; Federated to Acquire Rival Macy."
50. Lena H. Sun, "Retailing's High-Tech Revolution," *Washington Post,* February 12, 1989, H1.
51. "Industry Snapshot."
52. Joseph Maglitta, "Centralized, Revitalized IS Brings Sears into the '90s," *Computerworld,* October 8, 1990, 47; Kara Swisher, "The (Lowest) Price Is Right," *Washington Post,* December 21, 1992, F1.
53. Bradley Johnson, "Supermarkets Take 'Position,'" *Advertising Age,* May 10, 1993, S1.
54. Interview with Willard "Bill" Bishop, September 8, 2014.
55. Gary Levin, "Marketers Flock to Loyalty Offers; Programs Acknowledging the Value of Keeping Your Best Customers," *Advertising Age,* May 24, 1993, 13.

56. Art Turock, [no title], *Progressive Grocer,* May 1, 2003, via LexisNexis.
57. Scott Sandberg, quoted in Judith Graham, "Stores See Loyal Customers Slip Away," *Advertising Age,* July 11, 1988, 12.
58. Bob Cappelli, quoted in Graham, "Stores See Loyal Customers Slip Away."
59. Janet Simons, "Albertsons' Discount Strategy Shakes Up Market," *Advertising Age,* April 28, 1986, S30.
60. Levin, "Marketers Flock to Loyalty Offers."
61. Frederick Reichheld, quoted in ibid.
62. Thomas Hoffman and Mitch Wagner, "Visions of Holiday $ugar-plums," *Computerworld,* December 4, 1995, 1.
63. John Groman, quoted in Levin, "Marketers Flock to Loyalty Offers."
64. Hoffman and Wagner, "Visions of Holiday $ugarplums."
65. Associated Press, "Retail Federation Posts Guidelines to Ensure Privacy," *Portland Press Herald,* April 15, 1998, 6C, via LexisNexis.
66. Amy Meyers Jaffe, "Retailing's New Strategy: I Can Get It for You Personal," *New York Times,* 3/8.
67. N. R. Kleinfeld, "Targeting the Grocery Shopper," *New York Times,* May 26, 1991, http://www.nytimes.com/1991/05/26/business/targeting-the-grocery-shopper.html?src=pm&pagewanted=3&pagewanted=all, accessed January 5, 2014.
68. See, for example, Riccardo Davis, "Shoppers, See the Special on Aisle 11's Floor," *Advertising Age,* January 25, 1993, 8; Robert Viney, "Solving the Agency-Client-Mismatch," *Advertising Age,* May 31, 1993, 20. In the mid-2000s, 70 percent was the number suggested. See Rob Holston, "Avoid the Shopper-Marketing Pitfalls," *Advertising Age,* March 31, 2008, 20.
69. Jack Neff, "Dina Howell; Director-First Moment of Truth Center for Expertise, Procter & Gamble Co.," *Advertising Age,* June 7, 2004, S12.
70. Ibid.
71. Emily DeNitto, "Catalina Effort Streams Coupon Process," *Advertising Age,* December 13, 1993, 25.
72. Kate Fitzgerald, "P&G Zero-Coupon Move Sparks Related Cutbacks by Competitors," *Advertising Age,* March 18, 1996, 3.

73. "Smaller Role for Coupons?" *Advertising Age,* January 22, 1996, 16.

74. DeNitto, "Catalina Effort Streams Coupon Process."

75. Alison Fahey, "For Grocery Stores, 'Keep It Simple,'" *Advertising Age,* August 27, 1990, S8.

76. Alison Fahey, "Advertising Media Crowd into Aisles," *Advertising Age,* June 18, 1990, via LexisNexis.

77. Laurie Freeman, "Supermarkets Sift Through Data; Work Needed to Take These Card Programs to the Next Level," *Advertising Age,* October 10, 1994, S16.

78. Kleinfield, "Targeting the Grocery Shopper."

79. Ibid.

80. Bradley Johnson, "Grocers Learn to Nibble, Not Gulp," *Advertising Age* (Special Report on Database Marketing,) January 13, 1992, 28.

81. Amanda Beeler, "Online Coupon Site Links with Fun and Games," *Advertising Age,* October 25, 1999, 48; Tommy Greer, "In-Store Incentives Work for All," *Advertising Age,* February 7, 1994, 30.

82. Beeler, "Online Coupon Site Links with Fun and Games."

83. Greer, "In-Store Incentives Work for All."

84. Bradley Johnson, "Grocers Learn to Nibble, Not Gulp," *Advertising Age,* January 13, 1992, 28.

85. Freeman, "Supermarkets Sift Through Data; Work Needed to Take These Card Programs to the Next Level."

86. Rick Barlow, president of Frequency Marketing, quoted in Freeman, "Supermarkets Sift Through Data; Work Needed to Take These Card Programs to the Next Level."

87. See Turow, *Niche Envy: Marketing Discrimination and the New Media World.*

88. Michael Schrage, "Out There: The Ultimate Network," *Adweek,* May 17, 1993, http://www.adweek.com/news/advertising/out-there-ultimate-network-bby-michael-schragbbr-clearnonebr-clearnonelooking-next-, accessed July 19, 2013.

89. From Joseph Turow and Ashkan Soltani, "Marketing and Selling in the Digital Age," unpublished paper for the Ohio State Attorney General, August 1, 2013, 11. The quote is from Soltani's contribution to this paper.

90. Ibid.

91. See Joseph Turow, *The Daily You: How the New Advertising Industry Is Defining Your Identity and Your Worth* (New Haven: Yale University Press, 2011), 34–64.

92. Patricia B. Seybold, "Old Rules Apply in the Marketplace of the Future," *Computerworld,* December 5, 1994, 35.

93. "A Globe of Villages," *Progressive Grocer,* February 1, 1997.

94. Seybold, "Old Rules Apply in the Marketplace of the Future."

95. Alan Alper, "Caught Up in the Web," *Computerworld,* June 1, 1996, R5.

96. Lauren J. Flynn, "Malls and Stores Find New Outlets in Cyberspace," *New York Times,* December 5, 1996, C2.

97. Rajiv Chandrasekaran, "More Shoppers Are Buying Online," December 24, 1997, C1, via LexisNexis.

98. Sharon Machlis, "Macy's to Grow Online," *Computerworld,* July 28, 1998; Karen Padley, "Fingerhut Is Sold for $1.7 Billion," *St. Paul Pioneer Press,* February 12, 1999, 1E.

99. Peter Grant, "A New Line for Retailer," *New York Daily News,* February 12, 1999, 53.

100. Karen Padley, "Fingerhut Is Sold for $1.7 Billion," *St. Paul Pioneer Press,* February 12 1999, 1E.

101. Michael J. Pachuta, "Wal-Mart Ready to Hit Start with Its New E-Tail Strategy," *Investor's Business Daily,* October 22, 1999, A16.

102. Frances Katz, "Federated Launches Interactive Division," *Atlanta Journal and Constitution,* June 27, 1998, 1E, via LexisNexis.

103. "Voting with Their Pocketbooks," *Progressive Grocer,* February 1, 1998.

104. Michelle Slattala, "Turning Coupon Users from Clippers into Clickers," *New York Times,* April 1, 1999, G13.

105. "Food Lion Selects Bigfoot Interactive for E-mail CRM Solutions," *Progressive Grocer,* July 7, 2004.

106. "Inside Line," *Advertising Age,* June 5, 2000, 35.

107. Ibid.

108. Joseph Tarnowski, "Groundbreakers," *Progressive Grocer,* April 2007.

109. "Broadway Marketplace to Roll Out Enabled Loyalty Program," *Progressive Grocer,* September 20, 2009.

110. Tarnowski, "Groundbreakers."

111. Amanda Lenhart, "Teens and Mobile Phones over the Past Five Years," Pew Research Center, August 19, 2009, http://www.pewinternet.org/2009/08/19/teens-and-mobile-phones-over-the-past-five-years-pew-internet-looks-back, accessed March 22, 2016.

112. Sarah Radwanick, "5 Years Later: A Look Back at the Rise of the iPhone," June 29, 2012, https://www.comscore.com/ita/Insights/Blog/5-Years-Later-A-Look-Back-at-the-Rise-of-the-iPhone, accessed March 22, 2016.

113. Ibid.

114. Phil Lembert, "Send in the Clones—or Not," *Progressive Grocer,* August 1, 2005.

115. Ibid.

116. Christina Veiders, "E-Baby," *Supermarket News,* May 2, 2011.

117. Ibid.

118. Elliot Zwiebach, "Amazon Mulls Grocery Expansion," *Supermarket News,* January 31, 2011, 4.

119. Ibid.

120. Joseph Bonney, "Amazon's Supply Chain," *Journal of Commerce Online,* January 20, 2012, via LexisNexis.

121. Much, "Brick-and-Mortar Chains Like Macy's Battle vs. Amazon."

122. Ibid.

123. Donald Melanson, "ComScore Report Finds 42 percent of US Mobile Users Have Smartphones, Android at Nearly 50 Percent," Engadget, February 23, 2012, http://www.engadget.com/2012/02/23/comscore-report-finds-42-percent-of-us-mobile-users-have-smartph/, accessed January 5, 2014.

124. Much, "Brick-and-Mortar Chains Like Macy's Battle vs. Amazon."

125. Ibid.

126. Ibid.

4 HUNTING THE MOBILE SHOPPER

1. Laura Heller, "Will Ship from Store the Retailers' Last Mile Solution," FierceRetail, June 13, 2014, http://www.fierceretail.com/story/will-ship-store-be-retailers-last-mile-solution/2014-06-13, accessed February 21, 2015.

2. Laura Heller, "Amazon Prime Adds 10 Million Members, Breaks Records," *Fierce Retail*, December 26, 2014, http://www.fierceretail.com/story/amazon-prime-adds-10-million-members-breaks-records/2014-12-26, accessed February 15, 2015.
3. "Our Company" and "The History of Peapod," http://www.peapod.com/site/companyPages/our-company-overview.jsp, accessed February 27, 2015.
4. Annie Gasparro, "Kroger Isn't Afraid of Online Grocery Shopping," *Wall Street Journal*, October 30, 2013, http://blogs.wsj.com/corporate-intelligence/2013/10/30/kroger-isnt-afraid-of-online-grocery-shopping/, accessed February 27, 2015.
5. Annie Gasparro, "Kroger Agrees to Buy Online Vitamin Seller Vitacoast," *Wall Street Journal*, July 2, 2014, http://www.wsj.com/articles/kroger-plans-to-buy-online-vitamin-seller-vitacost-1404300842, accessed February 27, 2015; Sarah Mahoney, "Walmart Adds Uber," *Marketing Daily*, June 3, 2016, http://www.mediapost.com/publications/article/277271/walmart-adds-uber-target-leaves-curbside.html, accessed June 3, 2016.
6. Shelly Banjo, "Rampant Returns Plague E-Retailers," *Wall Street Journal*, December 22, 2013, http://www.wsj.com/articles/SB10001424052702304773104579270260683155216, accessed February 21, 2015.
7. "Despite Potential for Heavy Traffic and Long Checkout Lines, Shoppers Will Spend More of Their Holiday Budget In-Store Than Online," Deloitte, November 10, 2014, http://www2.deloitte.com/us/en/pages/about-deloitte/articles/press-releases/potential-for-heavy-traffic-and-long-checkout-lines-shoppers-spend-more-holiday-budget-in-store-than-online.html, accessed February 21, 2015.
8. "Amazon Locker," http://www.amazon.com/b/?_encoding=UTF8&node=6442600011#f6f8f9?ref_=acs_ux_hsb_3s_1_s_slideshowtes& ref=spkl_3_0_1976108042&ie=UTF8&pf_rd_m=ATVPDKIKX0DER&pf_rd_s=desktop-auto-sparkle&pf_rd_r=0ZDH4MS841YTWQXHY7AS&pf_rd_p=1976108042&pf_rd_t=301&pf_rd_i=amazon%20lock er%20location&qid=1424876566, accessed February 25, 2015. Walmart seems to have had somewhat more success in Toronto, Canada, with the locker idea, which encompassed forty-three stores in late 2014,

while Amazon had not placed lockers in Canada at all. See Hollie Shaw, "Walmart looks to 'Grab' Online Sales," *Windsor Star,* November 28, 2014, B5.

9. "Instacart Frequently Asked Questions," https://www.instacart.com/faq, accessed February 27, 2015; "About Google Express," https://support.google.com/shoppingexpress/answer/4561693?hl=en, accessed February 27, 2015; Sarah Perez, "eBay Expands Local Pilot Program Offering Same-Day Delivery, In-Store Pickup in Brooklyn," http://techcrunch.com/2014/12/16/ebay-expands-local-pilot-program-offering-same-day-delivery-in-store-pickup-in-brooklyn/, accessed February 27, 2015.

10. Chase Martin, "Connected in the Air," IOT Daily, August 21, 2016, http://www.mediapost.com/publications/article/282917/connected-in-the-air-drones-take-flight-forcnn-v.html, accessed September 1, 2016.

11. Caitlyn Bohannon, "Amazon Prime Now Raises Bar with One-Hour Delivery," mobile commerce daily, December 19, 2014, http://www.mobilecommercedaily.com/amazon-prime-now-raises-bar-with-prime-now-one-hour-shipping, accessed February 23, 2015; Eric Sass, "Amazon Offers One-Hour Delivery with Prime-Now App," *MediaPost,* December 18, 2014, http://www.mediapost.com/publications/article/240426/amazon-offers-one-hour-delivery-with-prime-now-app.html, accessed February 23, 2014.

12. Bohannon, "Amazon Prime Now Raises Bar with One-Hour Delivery."

13. Dave Clark, quoted in Sass, "Amazon Offers One-Hour Delivery with Prime-Now App."

14. Heather Somerville, "Walmart Expands Delivery," *San Jose Mercury News,* October 17, 2013; "Walmart Uses Test-and-Learn Approach to New Products," *Telegraph-Journal* (New Brunswick, Canada), April 18, 2014, B2.

15. Alexia Elejade-Ruiz, "Macy's, Bloomingdale's Adding Same-Day Delivery," *Chicago Tribune,* September 15, 2014, http://www.chicagotribune.com/business/breaking/chi-macys-bloomingdales-same-day-delivery-20140915-story.html, accessed February 25, 2015.

16. Jonathan Berr, "Macy's: Can Same-Day Delivery Deliver Results?" CBS News.com, February 24, 2015, http://www.cbsnews.com/news/macys-expands-successful-same-day-delivery-test/, accessed February 25, 2015.
17. Paul Weitzel, quoted in Sarah Mahoney, "Walmart Adds Uber," *Marketing Daily*, June 3, 2016, http://www.mediapost.com/publications/article/277271/walmart-adds-uber-target-leaves-curbside.html, accessed June 3, 2016.
18. Mary Meeker, "Internet Trends 2016," KPCB, June 1, 2016, http://www.kpcb.com/internet-trends, accessed June 11, 2016.
19. Justin Honaman, "Top 6 Retail Trends to Watch in 2014," RIS-Retail Info Systems News, http://risnews.edgl.com/retail-news/Top-6-Retail-Trends-to-Watch-in-201490590, January 10, 2014.
20. Ibid.
21. Ibid.
22. Ibid.
23. Puneet Mehta, "The Biggest Opportunity for CPG Brands on Mobile: A Killer Loyalty Program," *MediaPost*, December 29, 2014, http://www.mediapost.com/publications/article/240820/the-biggest-opportunity-for-cpg-brands-on-mobile.html, accessed February 15, 2015.
24. "Turnstile," Wikipedia, http://en.wikipedia.org/wiki/Turnstile, accessed March 9, 2015.
25. Christine Blank, "Retailers Face New Mobile Tracking Code of Conduct," FierceRetail, October 22, 2013, http://www.fierceretail.com/mobileretail/story/retailers-face-new-mobile-tracking-code-conduct/2013-10-22, accessed March 28, 2015.
26. Brigid Sweeney, "Attention, Shoppers: This Man Is Tracking You," *Crain's Chicago Business*, June 22, 2013, http://www.chicagobusiness.com/article/20130622/ISSUE01/306229976/attention-shoppers-this-man-is-tracking-you, accessed March 4, 2015.
27. "How Euclid Location Analytics Works," Euclid Analytics, http://euclidanalytics.com/products/technology/, accessed September 30, 2015. Also, "Retail Insights as Specialized as You," Euclid Analytics, http://euclidanalytics.com/solutions/retail/, accessed September 30, 2015.

28. Interview with Bill Martin, March 4, 2015.

29. Interview with anonymous ShopperTrak executive, March 4, 2015.

30. See also "Interior Location Analytics," ShopperTrak, https://vimeo.com/70339461?mkt_tok=3RkMMJWWfF9wsRonu6%2FMZKXonjHpfsX56u4tXKCwlMI%2F0ER3f OvrPUfGjI4ATMZrI%2BSLDwEYGJlv6SgFT7PDMbR00LgMWhM%3D, accessed March 9, 2015.

31. Martin Wooley, "Bluetooth Technology: Protecting Your Privacy," Bluetooth Special Interest Group, April 2, 2015, http://blog.bluetooth.com/bluetooth-technology-protecting-your-privacy/, accessed August 25, 2015; "Our History," Bluetooth Special Interest Group, http://www.bluetooth.com/Pages/History-of-Bluetooth.aspx, accessed August 25, 2015.

32. Peter Cohan, "How Nordstrom Uses Wifi to Spy on Shoppers," *Forbes*, May 9, 2013, http://www.forbes.com/sites/petercohan/2013/05/09/how-nordstrom-and-home-depot-use-wifi-to-spy-on-shoppers/, accessed March 7, 2015. See also Quentin Hardy, "Technology Turns to Tracking People Off-Line," *New York Times*, March 7, 2013, http://bits.blogs.nytimes.com/2013/03/07/technology-turns-to-tracking-people-offline/?_r=0, accessed March 7, 2015.

33. "Privacy Matters," Euclid, http://euclidanalytics.com/about/privacy/, accessed March 7, 2015.

34. Lisa Vaas, "Nordstrom Tracking Customer Movement via Smart Phones' Wi-Fi sniffing," nakedsecurity, May 9, 2013, https://nakedsecurity.sophos.com/2013/05/09/nordstrom-tracking-customer-smartphones-wifi-sniffing/, accessed March 8, 2015.

35. Kerry Wright, "Apple Randomize MAC Addresses with iOS 8 . . . Or Did They?," Purple Wi-Fi, October 7, 2014, http://www.purplewifi.net/apple-randomise-mac-addresses-ios-8/, accessed March 8, 2015.

36. Graham Smith, "Man in the Middle," LinkedIn, July 19, 2015, https://www.linkedin.com/pulse/man-middle-attacks-graham-smith-phd, accessed August 24, 2015.

37. Julian Bhardwaj, "What Is Your Phone Saying Behind Your Back," nakedsecurity, October 2, 2012, https://nakedsecurity.sophos.com/2012/10/02/what-is-your-phone-saying-behind-your-back/, accessed March 8, 2015.

38. Wright, "Apple Randomize MAC Addresses with iOS 8."
39. Interview with anonymous source, June 25, 2015.
40. Wright, "Apple Randomize MAC Addresses with iOS 8."
41. Jimmy Buchheim (CEO of StickNFind), in Clint Boulton, "Beacon Technology Could Pose Security Challenges, Vendor Says," *Wall Street Journal*, http://blogs.wsj.com/cio/2014/07/22/beacon-technology-could-pose-security-challenges-vendor-says/, accessed March 9, 2015.
42. Vincent Gao, "Everything You Always Wanted to Know About Bluetooth—Security in Bluetooth 4.2," Bluetooth Special Interest Group, http://blog.bluetooth.com/everything-you-always-wanted-to-know-about-bluetooth-security-in-bluetooth-4-2/?_ga=1.164912767.1332021922.1459354939, accessed August 24, 2015.
43. Interview with Raul Verano of Shopperception, March 27, 2015.
44. Jason Ankeny, "Innovator: shopkick's Cyriac Roeding Reinvents Retail," *Entrepreneur*, February 22, 2011, https://www.entrepreneur.com/article/218157, accessed May 3, 2016.
45. "Proximity Marketing."
46. "Germany's Largest Retailers Launch Nationwide Programs on Day One," PR Newswire, October 23, 2014.
47. "Shopkick Surpasses $1 Billion Generated for Partners," PR Newswire, October 15, 2014, http://www.prnewswire.com/news-releases/shopkick-expands-into-first-international-market-five-of-germanys-largest-retailers-launch-nationwide-programs-on-day-one-707143761.html, accessed March 12, 2015.
48. Christine Blank, "Macy's shopkick Test Confirms Value of iBeacon," FierceRetail, November 20, 2013, http://www.fierceretail.com/story/macy%E2%80%99s-shopkick-test-confirms-value-of-ibeaon; "shopkick Privacy Policy," https://www.shopkick.com/privacy-policy, accessed March 10, 2015.
49. "Shopkick Debuts shopBeacon," PR Newswire, November 20, 2013, http://www.prnewswire.com/news-releases/shopkick-debuts-shopbeacon-232652521.html, accessed June 30, 2016.
50. Lauren Johnson, "Hillshire Brands Sees 20% Jump in Purchase Intent with Beacons," *Adweek*, July 22, 2014, http://www.adweek.com/

news/technology/hillshire-brands-sees-20-jump-purchase-intent-beacons-159042, accessed April 30, 2016.

51. "Get Rich Insights into Shopper Behavior," inMarket, http://www.inmarket.com/retail.html#timingmatters, accessed March 26, 2015.

52. "Privacy Policy," inMarket, January 4, 2014 (effective date), https://www.inmarket.com/privacy_policy.html, accessed September 24, 2015.

53. Jordan Kahn, "Here's Why Beacon Networks Are the Way to Go for Retailers and App Developers Supporting iBeacon," 9TO5Mac, http://9to5mac.com/2014/11/19/comscore-ibeacon-app-usage/, accessed March 29, 2015.

54. Interview with Jeff Griffin.

55. In somewhat torturous language, the policy described collecting information "you choose to provide through the linking of devices where you are actively choosing to share personal information, including health information, to interact with an application or program we host on our site." "Online Privacy & Security Policy," Walgreens, http://www.walgreens.com/topic/generalhelp/privacyandsecurity.jsp, accessed March 30, 2015. See also Margi Murphy, "Walgreens: Apple May Have a New Watch—but Collecting Data from Wearables Is Already in Motion," Computerworld UK, September 11, 2014, http://www.computerworlduk.com/in-depth/it-business/3571685/walgreens-apple-may-have-new-watch—but-collecting-data-from-wearables-is-already-in-motion/?intcmp=in_article, accessed March 30, 2015.

56. "Four Ways Walmart Uses Analytics," SAS: The Power to Know, http://www.sas.com/en_us/insights/articles/analytics/four-ways-walmart-uses-analytics.html, accessed March 30, 2015.

57. "Target Privacy Policy," Target Stores, http://www.target.com/spot/privacy-policy#InfoCollected, and http://m.target.com/spot/terms/privacy-policy#InformationShared, accessed March 29, 2015.

58. Gregory Piatetsky, "Did Target Really Predict a Teen's Pregnancy? The Inside Story," KDnuggets, May 7, 2014, http://www.kdnuggets.com/2014/05/target-predict-teen-pregnancy-inside-story.html, accessed March 29, 2015.

59. "Privacy Policy," Kroger, https://www.kroger.com/topic/privacy-policy, accessed March 29, 2015.

60. Tom Groenfelt, "Kroger Knows Your Shopping Patterns Better Than You Do," *Forbes,* http://www.forbes.com/sites/tomgroenfeldt/2013/10/28/kroger-knows-your-shopping-patterns-better-than-you-do/, accessed March 29, 2015.

61. "SCAN IT! & SCAN IT! mobile," Stop & Shop, http://stopandshop.com/shopping/shopping-tools/scanit/, accessed March 30, 2015.

62. Laura Heller, "ShopRite Pilots Mobile Scan," FierceMobileRetail, February 4, 2014, http://www.fierceretail.com/story/shoprite-pilots-mobile-scan-app/2014-02-04; Chantal Tode, "Walmart Boosts Scan & Go Self-Checkout with Mobile Coupons," *Mobile Commerce Daily,* August 2, 2013, http://www.mobilecommercedaily.com/walmart-boosts-scan-go-self-checkout-with-mobile-coupons, accessed June 12, 2016.

63. "Catalina Mobile," Catalina Marketing, http://www.catalinamarketing.com/wp-content/uploads/2014/02/Catalina_Mobile_PS1.pdf, accessed March 30, 2015.

64. "Six Things Every Retailer Must Do to Win in Mobile Commerce," Catalina Marketing, 2013, 8, http://www.catalinamarketing.com/wp-content/uploads/2014/01/Six_Things_Study.pdf, accessed September 23, 2015.

65. Ibid.

66. Brian Feldt, "Beacon Technology-Related Retail Sales to Grow Tenfold by 2017," *St. Louis Business Journal,* June 16, 2015, http://www.bizjournals.com/stlouis/blog/biznext/2015/06/beacon-technology-related-retail-sales-to-grow.html, accessed August 10, 2015.

67. Comments by Lung Huang, head of Strategic Partnerships, 84.51 Degrees, at *MediaPost*'s IoT: Shopping conference, New York, August 6, 2015.

68. "Why You'll Love IndoorAtlas," IndoorAtlas, https://www.indooratlas.com/, accessed August 10, 2015.

69. Ibid.

70. "Indoor Positioning System," GE Lighting, http://www.gelighting.com/LightingWeb/na/solutions/control-systems/indoor-positioning-system.jsp, accessed April 30, 2016.

71. Jordan Kahn, "GE Integrates iBeacons in New LED Lighting Fixtures Rolling Out in Walmart & Other Retailers," 9TO5Mac, May 29, 2014, http://9to5mac.com/2014/05/29/ge-integrates-ibeacons-in-new-led-lighting-fixtures-rolling-out-in-walmart-other-retailers/, accessed March 30, 2015. The GE fixtures also incorporated a new form of tracking called VLC (visual light communication). Instead of the radio waves that characterize Wi-Fi and Bluetooth, it uses pulses of light sent to a smartphone camera to alert, and an app program to accept format. See Martin LaMonica, "Philips Creates Shopping Assistant with LEDs and Smart Phone," IEEE Spectrum, February 2014, http://spectrum.ieee.org/tech-talk/computing/networks/philips-creates-store-shopping-assistant-with-leds-and-smart-phone, accessed March 30, 2015.
72. "GPS Accuracy," GPS.gov, http://www.gps.gov/systems/gps/performance/accuracy/, accessed September 23, 2015.
73. "Who Is xAd," xAd, November 2014, http://www.xad.com/wp-content/uploads/xAd-Fact-Sheet_November-2014.pdf, accessed September 23, 2015.
74. "Search," xAd, http://www.xad.com/technology/#products, accessed March 14, 2015.
75. Discussion with Mike Boland, March 13, 2015.
76. Mike Boland, quoted in David Kaplan, "More Than Half of Mobile Ad Dollars Will Be Location-Based by 2018," Geomarketing, May 29, 2014, http://www.geomarketing.com/more-than-half-of-mobile-ad-dollars-will-be-location-based-by-2018, accessed March 12, 2015.
77. Love Hudson-Maggio, "What Is Geo-Conquesting, and How Can It Drive Campaign Results?," localsolutions, August 13, 2014, http://blog.cmglocalsolutions.com/what-is-geo-conquesting-and-how-can-it-drive-campaign-results, accessed April 8, 2015.
78. Presentation by Scott Varlard, technology expert from the IPG Media Lab, at *MediaPost*'s IoT Shopping conference, New York, August 6, 2015.
79. Presentation by David Skaff, co-founder and head of creative, The Science Project, at *MediaPost*'s IoT Shopping conference, New York, August 6, 2015.

80. Presentation by Leigh Christie, manager, IsobarNow Lab Americas, at *MediaPost*'s IoT Shopping conference, New York, August 6, 2015.
81. Interview with Jeff Griffin, executive vice president, inMarket, June 9, 2015.

5 LOYALTY AS BAIT

1. Ron Lieber, "Now May Be a Good Time to Bail Out of Frequent-Flyer Programs," *New York Times*, February 28, 2014, http://www.nytimes.com/2014/03/01/your-money/credit-and-debit-cards/now-may-be-a-good-time-to-bail-out-of-frequent-flier-programs.html?_r=0, accessed April 30, 2015.
2. Fred Reichheld, "Letter to the Editor," *Harvard Business Review*, November 2002, 126.
3. Interview with anonymous source, August 21, 2014.
4. Peter Sachse, quoted in Allison Schiff, "Macy's CMO Shares Loyalty Insights at NRF Show," Direct Marketing, January 16, 2012, http://www.dmnews.com/multichannel-marketing/macys-cmo-shares-loyalty-insights-at-nrf-big-show/article/223344/, accessed September 26, 2015.
5. Laura Heller, "Target's Transformation Roadmap Makes Mobile the Front Door," *FierceMobileIT*, April 9, 2015, http://www.fiercemobileit.com/story/targets-transformation-roadmap-makes-mobile-front-door/2015-04-09, accessed May 16, 2016.
6. Leonie Barrie, "NRF 2014: Tailoring Technology for Omnichannel Success," just-style.com, January 21, 2014, http://www.just-style.com/analysis/tailoring-technology-for-omnichannel-success_id120413.aspx.
7. Jody Sarno, with Carlton A. Doty and Collin Colburn, "The Age of the Customer Requires a More Intelligent Enterprise," Forrester Research, January 22, 2014, 1, https://www.forrester.com/The+Age+Of+The+Customer+Requires+A+More+Intelligent+Enterprise/fulltext/-/E-RES112082, accessed April 27, 2015.
8. Melissa Healy, "DNA Sequencer Raises Doctors' Hopes for Personalized Medicine," *Los Angeles Times*, January 3, 2014, http://articles.latimes.com/2014/jan/03/science/la-sci-personalized-medicine-20140104, accessed September 1, 2015.

9. Ibid.
10. "Frequently Asked Questions," ancestry.com, http://dna.ancestry.com/legal/faq, accessed September 1, 2015.
11. Hans Buhlmann, "The Actuary: The Role and Limitation of the Profession Since the Mid-19th Century," *ASTIN Bulletin* 27:2 (November 1997): 165–71.
12. Quoted in "Reinventing Society in the Wake of Big Data: A Conversation with Alex (Sandy) Pentland," Edge.org, https://edge.org/conversation/reinventing-society-in-the-wake-of-big-data, accessed September 26, 2015.
13. "Engaging the Selective Shopper," Catalina Marketing, 2013, 11, http://info.catalinamarketing.com/files/133/Engaging-the-Selective-Shopper-12-6-13.pdf, accessed April 30, 2015.
14. Lisa Tourville et al., "Actuaries in Advanced Business Analytics," Society of Actuaries, 9, https://www.soa.org/Files/Soa/act-adv-bus-analytics-paper.pdf, accessed April 27, 2015.
15. Ibid.
16. Stephanie Baghdassarian et al., "Predicts 2014: Apps, Personal Cloud and Data Analytics Will Drive New Customer Interactions," Gartner, Inc., November 22, 2013, 2, http://www.gartner.com/document/2628016?ref=solrResearch&refval=170082897&qid=96c3e252d55b6b8814dcb2f266e8a7e9, accessed June 30, 2016.
17. Ibid., 9.
18. Holman W. Jenkins, Jr., "Google and the Search for the Future," *Wall Street Journal*, August 14, 2010, http://www.wsj.com/articles/SB10001424052748704901104575423294099527212.
19. See Joseph Turow, *The Daily You: How the New Advertising Industry Is Defining Your Identity and Your Worth* (New Haven: Yale University Press, 2011), 65–87.
20. See, for example, Joseph Turow et al., *Americans, Marketers, and the Internet, 1999–2012* (Philadelphia: Annenberg School for Communication, 2015), http://repository.upenn.edu/asc_papers/348/.
21. "Identity-Driven Marketing: Best Practices for Marketing Continuity," Janrain, 2015, 9, http://janrain.com/resources/white-papers/identity-driven-marketing-best-practices-for-marketing-continuity/, accessed September 26, 2015.

22. Janrain, "Marketing Continuity: A Strategic Framework for Creating Connected Customer Experiences," http://www1.janrain.com/rs/janrain/images/White-Paper-Marketing-Continuity-Strategic-Framework-Creating-Connected-Customer-Experiences.pdf, 16, accessed September 26, 2015.
23. See Tapad website, http://www.tapad.com/product/#advertiser-solution, accessed September 26, 2015.
24. "Nielsen Confirms Tapad Cross Device Accuracy at 91.2%," Tapad Blog, December 2, 2014, http://www.tapad.com/nielsen-study-finds-tapads-device-connections-91-2-percent-accurate/, accessed September 26, 2015.
25. "Privacy Policy," Tapad.com, July 30, 2015, http://www.tapad.com/privacy-policy/, accessed September 2, 2015.
26. "Pinpoint advertising," Tapad.com, http://www.tapad.com/lifestyle/advertising/, accessed September 3, 2015.
27. "Identity-Driven Marketing: Best Practices for Marketing Continuity," 6.
28. Janrain, "Marketing Continuity," 15.
29. "Identity-Driven Marketing: Best Practices for Marketing Continuity," 6.
30. "Acxiom Disrupts Conventional Marketing Models with New Audience Operating System (AOS)," http://www.acxiom.com/acxiom-disrupts-conventional-marketing-models-with-new-audience-operating-system-aos/, accessed August 28, 2015.
31. Joe Mandese, "Supplier of the Year: Acxiom," *MediaPost,* January 8, 2014, http://www.mediapost.com/publications/article/216930/supplier-of-the-year-acxiom-whos-on-first-wha.html, accessed August 28, 2015.
32. Kathryn Zickuhr, "Location-Based Services," Pew Research Center, September 12, 2013, http://www.pewinternet.org/2013/09/12/location-based-services/, accessed September 26, 2015.
33. See, for example, Sonja Utz and Nicole Kramer, "The Privacy Paradox on Social Network Sites Revisited," *Cyberpsychology* 3, no. 2 (2009), http://cyberpsychology.eu/view.php?cisloclanku=2009111001%26article=1, accessed April 26, 2015; Susan B. Barnes, "A Privacy Paradox: Social Networking in the United States," First Monday, 11, http://firstmonday.org/htbin/cgiwrap/bin/ojs . . ., accessed August 8, 2008.

34. Brad Stone, "Our Paradoxical Attitudes Toward Privacy," *New York Times*, July 2, 2008, http://bits.blogs.nytimes.com/2008/07/02/our-paradoxical-attitudes-towards-privacy/, accessed April 26, 2015.
35. See "US Consumers Want More Personalized Retail Experience and Control over Personal Information, Accenture Survey Shows," Accenture, March 9, 2015, http://newsroom.accenture.com/news/us-consumers-want-more-personalized-retail-experience-and-control-over-personal-information-accenture-survey-shows.htm, accessed April 26, 2015; "Shoppers Conflicted on How Personal to Get," Accenture, March 2015, http://www.accenture.com/SiteCollectionDocuments/PDF/Accenture-Retail-Personalization-Survey-Infographic-March-2015.pdf, accessed April 26, 2015.
36. "US Consumers Want More Personalized Retail Experience and Control over Personal Information, Accenture Survey Shows."
37. Carlton A. Doty, "The Power Customer Context," Forrester Research, March 21, 2016 (updated from April 14, 2014), https://www.forrester.com/report/The+Power+Of+Customer+Context/-/E-RES114961, accessed June 30, 2016.
38. Lars Meyer-Waarden and Christophe Benavent, "Grocery Retail and Loyalty Program Effects: Self-Selection or Retail Purchase Behavior Change?" *Journal of the Academy of Marketing Sciences* (2009) 37: 345–58.
39. Frederick F. Reichheld and Phil Schefter, "Your Secret Weapon on the Web," *Harvard Business Review*, July–August 2000, 113.
40. Jim Tierney, "Emotional Loyalty Emerges Among Many Brands, Customers," loyalty 360.org, October 2, 2013, http://loyalty360.org/Content-Gallery/Daily-News/Emotional-Loyalty-Emerges-Among-Many-Brands,-Custo.
41. "The Loyalty Report 2014 from Bond Brand Loyalty," Bond Brand Loyalty, 2014, http://cdn2.hubspot.net/hub/352767/file-942578460-pdf/Whitepapers/Bond_Brand_Loyalty_2014_Loyalty_Report_US.pdf?t=1430494118043, 2, accessed May 1, 2015.
42. Ibid.
43. Ayan Sen, "Gamifying Customer Engagement to Drive Growth," TeleTech, April 2015, http://www.teletech.com/thought-leadership/

articles/gamifying-customer-engagement-drive-growth#.VZ70qPlVhBc, accessed July 9, 2015.

44. Kristina Knight, "How to Use Big Data, Gamification for Better Rewards Programs," Bizreport, April 14, 2014, http://www.bizreport.com/2014/04/how-to-use-big-data-gamification-for-better-rewards-programs.html, accessed September 2, 2015.

45. Janrain, "Marketing Continuity," 16, accessed August 17, 2015.

46. "Identity-Driven Marketing," 8, accessed August 17, 2015.

47. Daniel Hofkin of William Blair, quoted in Gary M. Stern, "Making a Beauty Chain Pop Against All Rivals," *Investor's Business Daily,* September 24, 2012, A8.

48. Jason Gere, quoted by Ciaran McEvoy, "How to Pin Down a Profitable Niche Specialty," *Investors Business Daily,* December 15, 2014, A8.

49. "Specialty Chains Earnings Spotlight Holiday Trends," *Investor's Business Daily,* November 11, 2012, http://www.investors.com/news/ulta-movado-and-tiffany-report-earnings-this-week/, accessed May 19, 2016; and Helen Martin, "How Ulta Is Remaking the Customer Experience," FierceRetail, March 31, 2015, http://www.fierceretail.com/story/how-ulta-remaking-customer-experience/2015-03-31, accessed May 19, 2016.

50. See "Watch and Shop," UltaHaul Videos, http://www.ulta.com/getInspiredLanding.html?bd=true, accessed September 27, 2015.

51. Heather Martin, "How Ulta Is Remaking the Customer Experience," FierceRetail, March 31, 2015, http://www.fierceretail.com/story/how-ulta-remaking-customer-experience/2015-03-31, accessed May 7, 2015.

52. Heather Martin, "How Ulta Is Using Technology to Deliver a Distinctive Experience," FierceRetail, March 31, 2015, http://www.fierceretail.com/story/how-ulta-using-technology-deliver-distinctive-experience/2015-03-31, accessed May 19, 2016.

53. Ibid.

54. Nicole Marie Melton, "Sephora: We Want Customers 'Showrooming' in Our Stores," FierceRetail, January 13, 2014, http://www.fierceretail.com/story/sephora-we-want-customers-showrooming-our-stores/2014-01-13, accessed May 20, 2015.

55. Sen, "Gamifying Customer Engagement to Drive Growth."

56. "Sephora to Go," Mixrank, https://mixrank.com/appstore/apps/393328150/versions, accessed May 20, 2016.

57. Jefferson Graham, "Shopkick App Can Lead You to Discounts and Sales," *USA Today*, August 11, 2010, http://usatoday30.usatoday.com/tech/products/2010-08-12-shopkick12_ST_N.htm, accessed May 20, 2016.

58. "Move Your Business Forward by Streamlining Back-Office Operations," Sam's Club Member Services, https://www.samsclubms.com/cloveroffer/?utm_campaign=sams-club-clover-branded&utm_source=google&utm_medium=cpc&utm_term=%2Bperka&utm_content=perka&trkid=V1ADW271 094-20148910940-k-%2Bperka-64 058616860-b-s, accessed March 29, 2015.

59. "The Perka Privacy Statement, Last Updated August 20, 2014," http://perka.com/privacy/, accessed March 29, 2015.

60. Ibid.

61. Interview with anonymous source, June 25, 2015.

62. Laura Heller, "Savings Catcher Boosts Walmart App to Number One," FierceMobileRetail, August 7, 2014, http://www.fierceretail.com/mobileretail/story/savings-catcher-boosts-walmart-app-no-1/2014-08-07, accessed May 21, 2015.

63. "Target Privacy Policy," http://cartwheel.target.com/privacy-policy, accessed May 24, 2015.

64. Jillian Buttecali, "The Top 4 Omni-Channel Retailers," ID.me, January 15, 2015, https://blog.id.me/top-10-retail-onmi-channel-success-stories/, accessed May 20, 2015.

65. "Target Mobile App Privacy Policy," http://m.target.com/spot/terms/privacy-policy-app, accessed May 24, 2015.

66. "Target Privacy Policy."

67. "Shopkick Privacy Policy," https://www.shopkick.com/privacy-policy?utm_expid=31844936-2.DCxBLb4gTsm0wwWEELtEdA.0&utm_referrer=https:%2F%2Fwww.google.com%2F, accessed September 4, 2015.

68. "ULTA.com Privacy Policy, Updated July 31, 2015," https://www.ulta.com/ulta/common/privacyPolicy.jsp, accessed September 4, 2015.

69. "Safeway Privacy Policy (United States)," http://www.safeway.com/ShopStores/Privacy-Policy.page, accessed September 4, 2015.

70. See, for example, Thomas Husson, *The Future of Mobile Wallets Lies Beyond Payments,* Forrest Research, February 9, 2015, 11–15.

71. Ibid.

72. Interview with anonymous Vibes executive, June 27, 2015.

73. "About Passbook," Apple's iOS Developer Library, https://developer.apple.com/library/ios/documentation/UserExperience/Conceptual/PassKit_PG/Chapters/Introducti on.html, accessed September 1, 2015.

74. Interview with Vibes executive, August 31, 2015. Department store and supermarket chains got around the problem by asking customers to use the supermarkets' own sites or apps to save the coupons to their app or loyalty account. However, the shopper might not remember to buy the product linked to the discount coupon, so the store might still find the wallet useful as a location for a message that would become a location-appropriate lock screen reminder to purchase the product. Shoppers finding coupons online (from Coupons.com), for example, had to print them out and bring them to the score to be scanned—or input manually by the store clerk.

75. Michelle Saettler, "How Merchants Tie Loyalty to Mobile Wallets and Drive Results," Mobile Commerce Data, July 16, 2014, http://www.mobilecommercedaily.com/how-merchants-tie-loyalty-to-mobile-wallets-and-drive-results, accessed August 18, 2015.

76. Ibid.

77. Dave Richards, quoted in "U.S. Retailers Struggling to Meet Consumer Expectations Around Mobile and In-Store Experience, Accenture Study Finds," Accenture, March 25, 2015, https://newsroom.accenture.com/industries/retail/us-retailers-struggling-to-meet-consumer-expectations-around-mobile-and-in-store-experience-accenture-study-finds.htm.

78. Laura Heller, "Target's Transformation Roadmap Makes Mobile the Front Door," April 9, 2015, http://www.fiercemobileit.com/story/targets-transformation-roadmap-makes-mobile-front-door/2015-04-09, accessed September 28.

6 PERSONALIZING THE AISLES

1. *An A–Z Glossary of Personalized Marketing*, Neustar, 2014, 2, https://www.neustar.biz/resources/whitepapers/personalized-marketing-glossary, accessed June 30, 2016.
2. "Marketing Continuity: A Strategic Framework for Creating Connected Customer Experiences," 2014, http://www.janrain.com/resources/white-papers/marketing-continuity-strategic-framework-creating-connected-customer-experiences/ (page 5), accessed June 6, 2016.
3. Ibid. (page 19).
4. "Identity-Driven Marketing: Best Practices for Marketing Continuity," Janrain, 2015, http://www.janrain.com/resources/white-papers/identity-driven-marketing-best-practices-for-marketing-continuity/ (page 5), accessed June 6, 2016.
5. Ibid. (page 8).
6. Ibid. (page 14).
7. Ibid.
8. Ibid.
9. "Marketing Continuity" (page 14).
10. Ibid. (page 16).
11. Ibid.
12. "Offline Personalization Matters Just as Much," eMarketer, January 14, 2015, http://www.emarketer.com/Article/Offline-Personalization-Matters-Just-Much/1011815.
13. Kristen Sprague, "Keeping the E-Wolf at Bay," Furniture Today, November 4, 2014, http://www.furnituretoday.com/blogpost/13245-keeping-e-wolf-bay, accessed July 13, 2015.
14. Andrew Corselli, "Marketers Are Showing Customers Love," *DM News*, April 2015.
15. Adam Silverman, "Analyze This: Web Style Analytics Enters the Retail Store," Forrester Research, April 16, 2014, 1–2.
16. Mark Walsh, "Can Beacons Bring Home the Bacon?" *MediaPost*, August 19, 2014, http://www.mediapost.com/publications/article/232410/can-beacons-bring-home-the-bacon.html, accessed July 16, 2015.
17. Chuck Martin, "Beacon Data Tells When a Consumer Is Due to Shop," IOT Daily, November 10, 2015, http://www.mediapost.com/

publications/article/262237/beacon-data-tells-when-a-consumer-is-due-to-shop.html, accessed November 20, 2015.

18. Chuck Martin, "Beacons, GPS, Wi-Fi Combo: The New Mobile Presence," *MediaPost,* June 5, 2016, http://www.mediapost.com/publications/article/277300/beacons-gps-wi-fi-combo-the-new-mobile-presence.html, accessed June 5, 2016.

19. Ryan Joe, "Alliance Data Buys Epsilon a $2.3B Present: Conversant," Adexchanger, September 11, 2014, http://adexchanger.com/platforms/alliance-data-buys-epislon-a-2-3b-present-conversant/, accessed July 19, 2015.

20. "Our Approach," Conversant, http://www.conversantmedia.com/our-approach#identification_delivery, accessed July 19, 2015.

21. "Services," Agility/Harmony, http://www.agilityharmony.com/services, accessed July 19, 2015.

7 WHAT NOW?

1. "The Wearable Future," PricewaterhouseCoopers, http://www.pwc.com/us/en/technology/publications/wearable-technology.html, 42.

2. PricewaterhouseCoopers, "Wearable Technology Future Is Ripe for Growth—Most Notably Among Millennials, Says PwC US," October 21, 2014, http://www.pwc.com/us/en/press-releases/2014/wearable-technology-future.html, accessed October 8, 2015.

3. Broede Carmody, "iRing: Could Apple Soon Release a Smartring?," smartcompany, October 7, 2015, http://www.smartcompany.com.au/technology/48648-iring-could-apple-soon-release-a-smartring.html, accessed October 11, 2015.

4. Anick Jesdanun, "Fitness Tracker Sales Are High, But So Are Abandonment Rates," July 9, 2015, http://www.huffingtonpost.com/entry/fitness-tracker-sales-are-high-but-so-are-abandonment-rates_559e9cece4b05b1d028fd99b, accessed October 10, 2015.

5. Thomas Mulier, "Apple Helps Push U.S. Watch Sales to Biggest Drop in Seven Years," Bloomberg Business, August 7, 2015, http://www.bloomberg.com/news/articles/2015-08-07/apple-helps-push-u-s-watch-sales-to-biggest-drop-in-seven-years, accessed October 10, 2015.

6. Corinne Gretler, "Swatch Co-Inventor: Apple Will Succeed and an Ice Age Is Coming for Swiss Watches," Bloomberg Business, March 10, 2015, http://www.bloomberg.com/news/articles/2015-03-10/swatch-co-inventor-sees-apple-causing-ice-age-for-swiss-watches, accessed October 10, 2015.
7. "The Wearable Future." 11.
8. "A Look Ahead: How Wearable Technology Will Impact Retail," Certona, http://www.certona.com/a-look-ahead-how-wearable-technology-will-impact-retail/.
9. Ibid.
10. Ibid.
11. "The Wearable Future," 35, accessed October 10, 2015.
12. Nielsen, "Mobile Millennials: Over 85% of Generation Y Owns Smartphones," Snewswire, September 9, 2014, http://www.nielsen.com/us/en/insights/news/2014/mobile-millennials-over-85-percent-of-generation-y-owns-smartphones.html, accessed June 23, 2016.
13. Chloe Green, "The Next Customer Frontier: How Wearables Are Going Mainstream in Retail," *Information Age*, June 4, 2015, http://www.information-age.com/it-management/strategy-and-innovation/123459594/next-customer-frontier-how-wearables-are-going-mainstream-retail, accessed June 24, 2016.
14. The Wearable Future," 35, accessed October 10, 2015.
15. The Wearable Future," 5, accessed October 11, 2015.
16. "A Look Ahead: How Wearable Technology Will Impact Retail."
17. John Gillan, quoted in Allan Blair, "Is Now the Time for Retailers to Invest in Wearable Tech?," *Guardian*, April 1, 2015, http://www.theguardian.com/media-network/2015/apr/01/retailers-invest-wearable-tech, accessed October 10, 2015.
18. "The Wearable Future," 22.
19. Natasha Singer, "When No One Is Just a Face in a Crowd," *New York Times*, February 1, 2014, http://www.nytimes.com/2014/02/02/technology/when-no-one-is-just-a-face-in-the-crowd.html?_r=0, accessed October 14, 2015.
20. Ibid.

21. "Measurement and Analysis for Digital Signage," Retail Customer Experience/Networld Alliance, 2009, 6, http://www.retailcustomerexperience.com/static_media/filer_public/4c/24/4c24453f-6df5-486a-bd73-48a69f8aaa0b/ns_g_measurement-and-analysis.pdf, accessed October 14, 2015.
22. Larry Dignan, "The Future of Shopping: Where Psychology and Emotion Meet Analytics," ZDNet, January 15, 2014, http://www.zdnet.com/article/the-future-of-shopping-when-psychology-and-emotion-meet-analytics/, accessed October 12, 2015.
23. Paul Rubens, "Facial Recognition: Shop Where Everybody Knows Your Name," BBC online, December 9, 2014, http://www.bbc.com/news/business-30219820, accessed October 12, 2015.
24. Michael Casey, "Facial Recognition Software Is Scanning You Where You Least Expect It," CBS online, June 25, 2015, http://www.cbsnews.com/news/facial-recognition-software-is-scanning-you-where-you-least-expect-it/, October 12, 2015.
25. Rubens, "Facial Recognition."
26. Elizabeth Dwoskin and Evelyn M. Rusli, "The Technology That Unmasks Your Hidden Emotions," *Wall Street Journal,* January 28, 2015, http://www.wsj.com/articles/startups-see-your-face-unmask-your-emotions-1422472398, accessed October 12, 2015.
27. "Retail: Does Your In-Store Experience Bring Joy or Pain," Emotient, http://www.wsj.com/articles/startups-see-your-face-unmask-your-emotions-1422472398, accessed October 12, 2015. When Apple purchased Emotient, the firm "removed details [from its website] about the services it had been selling." See Rolf Winkler et al., "Apple Buys Artificial Intelligence Startup Emotient," January 7, 2016, http://www.wsj.com/articles/apple-buys-artificial-intelligence-startup-emotient-1452188715, accessed June 24, 2016. The original language quoted here is reproduced in Brian Rommele, "Apple's New AI Will Decode the 43 Muscles in Your Face and Help Siri2 Understand You Better," Medium.com, https://medium.com/@brianroemmele/apples-ai-secret-how-the-43-muscles-in-your-face-will-play-a-big-part-of-apple-s-new-ai-and-siri2-de8d4f54894#.o3vvwdrrd, accessed June 25, 2016.

28. Rommele, "Apple's New AI Will Decode the 43 Muscles in Your Face and Help Siri2 Understand You Better."

29. Dwoskin and Rusli, "The Technology That Unmasks Your Hidden Emotions."

30. Mario Trujillo, "Facial Recognition Quietly Taking Hold," *The Hill*, August 1, 2015, http://thehill.com/policy/technology/249958-facial-recognition-quietly-taking-hold, accessed October 12, 2015.

31. Rubens, "Facial Recognition: Shop Where Everybody Knows Your Name."

32. "VIP Identification," NEC, http://nl.nec.com/nl_NL/emea/solutions_services/it_solutions/security/vip.html, accessed October 14, 2015.

33. "Retail," FaceFirst, http://www.facefirst.com/services/retail, accessed October 14, 2015.

34. Rubens, "Facial Recognition: Shop Where Everybody Knows Your Name."

35. Ibid.

36. Pam Dixon, "The One-Way Mirror Society," World Privacy Forum, January 27, 2010, 35 and 37, http://www.worldprivacyforum.org/wp-content/uploads/2013/01/onewaymirrorsocietyfs.pdf, accessed October 14, 2015.

37. Casey, "Facial Recognition Software Is Scanning You Where You Least Expect It." See also Wendy Davis, "Illinois Privacy Law Tested by 'Faceprint' Cases," *Daily Online Examiner*, August 6, 2015, http://www.mediapost.com/publications/article/255620/illinois-privacy-law-tested-by-faceprint-cases.html, accessed June 26, 2016.

38. Christopher Matthews, "Private Eyes: Are Retailers Watching Our Every Move?" *Time*, September 18, 2012, http://business.time.com/2012/09/18/private-eyes-are-retailers-watching-our-every-move/, accessed September 18, 2012.

39. Casey, "Facial Recognition Software Is Scanning You Where You Least Expect It."

40. Ibid.

41. Ibid.

42. Trujillo, "Facial Recognition Quietly Taking Hold."

43. Ibid.

44. Chris Riddell, "The Future of Retail-Emotional Analytics," June 3, 2015, http://chrisriddell.com/the-future-of-retail-emotional-analytics/, accessed October 12, 2015.

45. Laurie Sullivan, "Zoomkube Introduces Facial Recognition in Retail Stores," Online Media Daily, August 23, 2013, http://www.mediapost.com/publications/article/207520/zoomkube-introduces-facial-recognition-in-retail-s.html, accessed May 16, 2014; Evan Schuman, "As Chain Trials Facial Recognition, Channel Assumptions Flip," StoreFrontBackTalk, July 1, 2013, http://archives.thecontent-firm.com/securityfraud/as-russian-convenience-store-chain-trials-facial-recognition-channel-assumptions-flip/, accessed April 18, 2014.

46. Schuman, "As Chain Trials Facial Recognition, Channel Assumptions Flip."

47. Ibid.

48. Ibid.

49. Ibid.

50. Ibid.

51. Ibid.

52. Quoted in Joseph Turow, *Niche Envy: Marketing Discrimination in the New Media World* (Cambridge, MA: MIT Press, 2006), 145.

53. "The Wearable Future," 22.

54. Alex Romanov, quoted in Alicia Fiorletta, "Will Wearable Technology Shake Up the Retail Landscape?," http://www.retailtouchpoints.com/features/trend-watch/will-wearable-technology-shake-up-the-retail-landscape, accessed October 11, 2015.

55. "The Wearable Future," 46.

56. Quoted in Turow, *Niche Envy: Marketing Discrimination in the New Media World,* 145.

57. In late 2015 Google was testing its ability to stream apps that people find through its search engine. It was part of an initiative Google called "app-first content," where search query results return some details from within apps served on the Google engine. See Laurie Sullivan, "Google Streams Apps Without Downloading, Serves Content in Search Results," SearchMarketingDaily, November 19, 2015, http://www.mediapost.com/publications/article/263020/

google-streams-apps-without-downloading-serves-co.html, accessed November 29, 2015.

58. See, for example, Chad Stone et al., "A Guide to Statistics on Historical Trends in Income Inequality," Center on Budget and Policy Priorities," http://www.cbpp.org/topics/inequality-trends, accessed October 19, 2015; Paul Krugman, "Why We're in a New Gilded Age," *New York Review of Books,* May 8, 2014, http://www.nybooks.com/articles/archives/2014/may/08/thomas-piketty-new-gilded-age/, accessed October 19, 2015.

59. Colin Gordon, "Growing Apart: A Political History of American Inequality," Center for Economic and Policy Research, August 26, 2013, http://scalar.usc.edu/works/growing-apart-a-political-history-of-american-inequality/index, accessed October 19, 2015.

60. http://pic.dhe.ibm.com/infocenter/initiate/v9r7/index.jsp?topic=%2Fcom.ibm.initiateglossary.doc%2Ftopics%2Fr_glossary_enterprise_customer_identity_management.html accessed April 7, 2014.

61. Pam Baker, "Shoppers OK with Online Tracking, Not So Much with In-Store Tracking," FierceRetailIT, July 15, 2013, http://www.fiercebigdata.com/story/shoppers-ok-online-tracking-not-so-much-store-tracking/2013-07-15, accessed October 20, 2015.

62. Yahoo! Advertising, *The Balancing Act: Getting Personalization Right,* May 2014, 11, https://advertising.yahoo.com/Insights/BALANCING-ACT.html, accessed May 8, 2015.

63. Brad Stone, "Our Paradoxical Attitudes Toward Privacy," *New York Times,* July 2, 2008, http://bits.blogs.nytimes.com/2008/07/02/our-paradoxical-attitudes-towards-privacy/, accessed April 26, 2015.

64. "The Truth About Shopping," McCann Truth Central, August 20, 2014, http://truthcentral.mccann.com/wp-content/uploads/2014/09/McCann_Truth_About_Shopping_Guide.pdf, accessed May 8, 2015.

65. Chuck Martin, "What the Shopper Gets Out of Being Tracked," *mCommerceDaily,* May 28, 2014, http://www.mediapost.com/publications/article/226734/what-the-shopper-gets-out-of-being-traked.html, accessed May 8, 2015.

66. Scott Snyder, "Mobile Devices: Facing the 'Privacy vs. Benefit' Trade-Off," Forbes.com, August 3, 2012, http://www.forbes.com/sites/ciocentral/2012/08/03/mobile-devices-facing-the-privacy-vs-benefit-trade-off/, accessed May 8, 2015.

67. Yahoo! Advertising, *The Balancing Act: Getting Personalization Right.*

68. "US Consumers Want More Personalized Retail Experience and Control over Personal Information, Accenture Survey Shows," Accenture, March 9, 2015, http://newsroom.accenture.com/news/us-consumers-want-more-personalized-retail-experience-and-control-over-personal-information-accenture-survey-shows.htm, accessed April 26, 2015.

69. See Joseph Turow et al., *Americans, Marketers, and the Internet, 1999–2012* (Philadelphia: Annenberg School for Communication, 2014), http://papers.ssrn.com/sol3/papers.cfm?abstract_id=2423753, accessed November 29, 2015. Chris Hoofnagle and Jennifer King of the Berkeley School of Law helped create the 2009 survey.

70. Google definition of resignation, https://www.google.com/?gws_rd=ssl#q=resignation, accessed May 18, 2015.

71. Chris Smith, "Tesco CIOL Personalisation Is the Next Big Thing for Retail Technology," *Guardian,* November 7, 2012, http://www.theguardian.com/media-network/2012/nov/07/tesco-retail-personalisation-technology, accessed November 29, 2015.

72. "The Future of Grocery," Nielsen, April 2015, 4, http://www.nielsen.com/content/dam/nielsenglobal/vn/docs/Reports/2015/Nielsen%20Global%20E-Commerce%20and%20The%20New%20Retail%20Report%20APRIL%202015%20(Digital).pdf, accessed November 29, 2015.

73. Scott Howe, "A Price Call-to-Action for the Data Industry," *Advertising Age,* April 8, 2014, http://adage.com/article/privacy-and-regulation/a-privacy-call-action-data-industry/292464/, accessed May 18, 2015.

74. Dixon, "The One-Way Mirror Society," 34.

75. Pam Baker, "Big Data Lobbyist Says Congress Won't Pass a Law to Protect Consumer Privacy," FierceBigData, August 20, 2014, http://www.fiercebigdata.com/story/big-data-lobbyist-says-congress-wont-pass-law-protect-consumer-privacy/2014-08-20, accessed October 19, 2015.

76. Robert Gellman, "Fair Information Practices: A Basic History, Version 2.13," February 11, 2015, http://bobgellman.com/rg-docs/rg-FIPShistory.pdf, accessed October 21, 2015.

77. Federal Trade Commission, *Protecting Consumer Privacy in an Era of Rapid Change,* December 2010, https://www.ftc.gov/sites/default/files/documents/reports/federal-trade-commission-bureau-consumer-protection-preliminary-ftc-staff-report-protecting-consumer/101201privacyreport.pdf, accessed May 16, 2015.

78. Helen Nissenbaum, "Respecting Context to Protect Privacy: Why Meaning Matters," *Sci. Eng. Ethics,* July 12, 2015.

79. Quoted in Natasha Singer, "Acxiom Lets Consumers See Details It Collects," *New York Times,* September 4, 2013, http://www.nytimes.com/2013/09/05/technology/acxiom-lets-consumers-see-data-it-collects.html, accessed November 30, 2015.

80. Singer, "Acxiom Lets Consumers See Details It Collects."

81. "Consumer Data Products Catalog," Acxiom, May 2013, 3, http://www.scribd.com/doc/187222488/Acxiom-Consumer-Data-Products-Catalog#scribd, accessed November 30, 2015.

82. See, for example, Anita Allen, *Unpopular Privacy* (New York: Oxford University Press, 1011); Kenneth Bamberger and Dierdre Mulligan, *Privacy on the Ground* (Cambridge, MA: MIT Press, 2015); Danielle Citron, *Hate Crimes in Cyberspace* (Cambridge, MA: Harvard University Press, 2016); Danielle Citron and Frank Pasquale, "The Scored Society," *Washington Law Review* 80:1 (2014); Julie Cohen, *Configuring the Networked Self* (New Haven: Yale University Press, 2013); Woodrow Hartzog and Frederic Stutzman, "The Case for Online Obscurity," *California Law Review* 101:1 (2013): 1; Chris Jay Hoofnagle, *Federal Trade Commission Privacy Law and Policy* (Cambridge: Cambridge University Press, 2016); Helen Nissenbaum, *Privacy in Context* (Stanford, CA: Stanford Law Books, 2009); Paul Ohm, "The Rise and Fall of Invasive Surveillance," *University of Illinois Law Review* 57 (2008): 1703–76; Frank Pasquale, *The Black Box Society* (Cambridge, MA: Harvard University Press, 2015); Joel Reidenberg, "Privacy in Public," *University of Miami Law Review* 69:141 (2014); Neil Richards, *Intellectual Privacy* (Oxford: Oxford University

Press, 2015); Marc Rotenberg, Jeramie Scott, and Julia Horwitz, eds., *Privacy in the Modern Age* (New York: New Press, 2015); Ira Rubinstein and Nathan Good, "Privacy by Design: A Counterfactual Analysis of Google and Facebook Privacy Incidents," *Berkeley Technology Law Journal* 28: 1333–1583; James Rule, *Privacy in Peril* (New York: Oxford University Press, 2009); Bruce Schneier, *David and Goliath* (New York: W.W. Norton. 2015); Daniel Solove, *Understanding Privacy* (Cambridge, MA: Harvard University Press, 2015); Peter Swire and Kenesa Ahmad, *U.S. Private Sector Privacy* (International Association of Privacy Professionals, 2012); Omer Tene and Jules Polonetsky, "A Theory of Creepy," *Yale Journal of Law and Technology* 59 (2013): 59–100; Tal Zarsky, "Transparent Predictions," *University of Illinois Law Review* 4 (2013): 1503–70.

INDEX